The Fitness of Information

The Fitness of Information

Quantitative Assessments of Critical Evidence

CHAOMEI CHEN

College of Computing and Informatics
Drexel University
Philadelphia, PA, USA

Library of Congress Cataloging-in-Publication Data:

Chen, Chaomei, 1960–
    The fitness of information : quantitative assessments of critical evidence / Chaomei Chen,
College of Computing and Informatics, Drexel University, Philadelphia, PA.
        pages   cm
    Includes bibliographical references and index.
    ISBN 978-1-118-12833-6 (cloth)
1. Social sciences–Research–Methodology.   2. Quantitative research.   3. Criticism.   I. Title.
    H62.C363 2014
    001.4′2–dc23
                                                                                    2014007306

Printed in the United States of America

ISBN: 9781118128336

10 9 8 7 6 5 4 3 2 1

Contents

Preface

"Big data" has become a buzzword in more and more countries.

We are all excited about the possibility that the massive computational power will solve some of the most challenging problems we have ever encountered. Something important, however, seems lost in this modern "gold rush" for big data. Now we have the ability to collect and store a wide variety of data at an unprecedented rate, and the rate grows much faster than what we can manage to digest even a tiny portion of the data universe. It is a good time for us to ask ourselves: why are we interested in gathering data in the first place? What are we looking for in a haystack of data or an ocean of data? What is the perceived value of a dataset? What is the additional value of a set of big data?

Many enthusiastic big data advocates have already given excellent answers to these questions. As an information scientist, my focus is on the outcome of data-driven paradigms, the quality of new solutions, and what we may still miss after everything is scaled up. It is my philosophy that the analyzer's mind-set matters to a great degree. What we will be able to discover is largely determined by what we set out to look for. It is in this context that I present this book to the readers.

This book addresses some of the questions that we often take for granted but are potentially crucial in how we process information. For example, what attracts our attention? What retains our interest? What are we trying to achieve by marshalling evidence? How would we know where we are in a big picture of an ongoing situation? How would we know if there are any alternative paths that may take us to the same destination? These are the kind of questions we will address in the seven chapters of this book through a fundamental theme—*the fitness of information*.

Chapter 1 focuses on answering a series of questions pertaining to the relationship between a source of information and our cognition and behavior, from factors that affect our attention to properties of beauty.

Chapter 2 explores the role of mental models in the analysis of information and in making decisions. We also emphasize the risk of having a wrong mental model and explain why we often miss signs that seem so obvious in hindsight.

Chapter 3 underlines the subjectivity of evidence through a series of examples of criminal trials. Our goal is to draw attention to the fact that the interpretation of evidence is determined by the purpose of an argument, which is in turn determined by the mental model of the interpreter.

Chapter 4 outlines the visual analytics framework that has been guiding the development of the CiteSpace system, which is designed for detecting emerging trends and identifying intellectual turning points. The workflow of the analytic process will illustrate why it is necessary for the analyst to examine evidence at different levels of granularity. Visualized changes in scientific knowledge will demonstrate the dynamics of knowledge as a complex adaptive system.

Chapter 5 reviews the origin of the fitness landscape metaphor in evolutionary genetics and subsequent expansions to other disciplines such as business and drug discovery. The fitness landscape metaphor helps us develop a generic framework for situation awareness and performance evaluation.

Chapter 6 consolidates the framework through a series of reviews of studies of radical and high-impact patents. This chapter examines the recurring theme of recombinant search and boundary spanning from a number of distinct but interconnected perspectives to characterize the role of recombinant search in generating radical and high-impact patents.

Chapter 7 provides an example of an analysis of portfolios of grant proposals and a few more examples of analyzing portfolios of publications. This chapter highlights the advantages of an interactive dual-map overlay design for a new type of portfolio analysis. Trajectories of various "populations" are visualized on an evolutionary landscape of knowledge to highlight the potential of the signs that may lead not only to hindsight but also to insight and foresight.

The fitness of information is written for anyone who would be interested in addressing these questions as part of analytic reasoning and decision-making. The focus is on explaining the essential principles and concepts to readers from a broad range of backgrounds. I expect that graduate students will be able to make most of use of the materials presented in this book, although others should find at least some of the chapters valuable for their needs. Many visualization and analytic results presented in the book are generated by our explanatory and computational systems, notably the CiteSpace system and its variations. The book covers the principles and analytic processes that drive the design and implementations of our tools. Thus, the book provides a good guidance to readers who would like to use our tools.

I have worked on several research projects in the past few years. These projects have provided a valuable source of inspiration, in particular, including a collaborative project sponsored by Pfizer on visualization and modeling of a lead compound optimization processes, a CiteSpace-guided systematic review of regenerative medicine for the *Expert Opinion on Biological Therapy* (EOBT), a research project on a study of clinical trials sponsored by IMS Health, and a gap analytics project funded as part of the NSF Center for Visual Decision and Informatics. I have also given numerous talks hosted by enthusiastic and visionary individuals from a diverse range of disciplinary backgrounds. I would like to take this opportunity to say thank you to my collaborators, colleagues, and friends: Loet Leydesdorff (University of Amsterdam, the Netherlands), Pak Chung Wong (PNNL, USA), Bruce A. Lefker (Pfizer, USA), Jared Milbank (Pfizer, USA), Mark Bernstein (Law, Drexel), James E.

Barrett (Pharmacology and Physiology, Drexel), Eileen Abels (Simmons College, USA), Rod Miller (College of Computing and Informatics, Drexel), Sue Clark (SunGard, USA), Paula Fearon (NIH, USA), Patricia Brennan (Thomson Reuters, USA), Dick Klavans (SciTech Strategies, USA), Eugene Garfield (Thomson Reuters, USA), Henry Small (SciTech Strategies, USA), and Emma Pettengale (EOBT, UK). I would also like to say thank you to my collaborators at Dalian University of Technology in China in association with my visiting professorship as a Changjiang Scholar; in particular, Zeyuan Liu, Yue Chen, Xianwen Wang, Zhigang Hu, and Shengbo Liu.

This book project would not be possible without the insightful initiative and encouragement at various stages from Jacqueline Palmieri, Steve Quigley, and Sari Friedman of John Wiley & Sons, Inc. I am grateful for their efforts and expertise in collecting insightful reviews and constructive comments from domain experts on initial plans and in helping me to keep the book project on track.

As always, I am grateful to all the support and understanding from the lovely members of my family, my wife Baohuan, and my sons Calvin and Steven, who have also offered numerous comments and suggestions on earlier drafts of the manuscript.

CHAOMEI CHEN

CHAPTER 1

Attention and Aesthetics

1.1 ATTENTION

Nothing in our universe can travel faster than the speed of light.

When scientists at the European Organization for Nuclear Research (CERN) found a beam of neutrinos that apparently traveled 60 billionths of a second faster than the speed of light in September 2011, they knew that their experiment, OPERA, would become either the most fundamental scientific breakthrough or one of the most widely publicized blunders.

Neutrinos rarely interact with normal matter. The beam of neutrinos traveled 454 miles through the crust of the Earth, from Geneva to Italy's Gran Sasso National Laboratory as if it were a vacuum. According to Albert Einstein's relativity theory, to make anything travel in a vacuum faster than light, it would take an infinite amount of energy. If the surprising result from the OPERA experiment turns out to be true, then the entire modern physics is at stake.

Scientists behind the OPERA were cautious. They made their data publicly available. Other scientists were cautious with their reactions too, although the news made headlines all over the world. Many agreed that such claims should be treated very carefully and validated in as many ways as possible.

In June 8, 2012, new results were presented by the research director of CERN's physics lab at the *25th International Conference on Neutrino Physics and Astrophysics* in Kyoto, Japan. The anomaly from the OPERA's original stunning measurement was inaccurate. It was an artifact of the measurement.

The question is what it takes to overturn a scientific theory. Did OPERA's claim ever really have a chance to undermine Einstein's theory? Did it make the theory of relativity stronger than before? Did journalists seem to be more excited than scientists in embracing the anomaly speed? Why would such news tend to attract a lot of attention from us in general?

The Fitness of Information: Quantitative Assessments of Critical Evidence, First Edition. Chaomei Chen.
© 2014 John Wiley & Sons, Inc. Published 2014 by John Wiley & Sons, Inc.

1.1.1   What Is It That Attracts Our Attention?

Have you ever noticed how it seems so much harder to distinguish people of a different race than from your own? And have you ever wondered how parents of twins were able to tell them apart? Researchers have noticed that babies seem to have a special ability that grown-ups no longer have. After nine months, the ability seems to be lost forever—the ability to recognize faces of other races and even other species, such as monkeys and gorillas. How do we know if an infant is interested in the face of a human or the face of a monkey? It is suggested that if an infant is staring at a human face, he is interested in the face and, furthermore, he is yet to fully understand the face. For example, if a picture of a monkey face and a human face are juxtaposed, the infant will stare at the face that is less familiar to him. To the infant, there appears something more to be discovered from the face that he is looking at. In contrast, if the infant is not paying much attention to the face of a human, but instead is turning his attention to the face of a monkey beside the human, then the interpretation is that the infant has already learned how to decode a human face, but he is interested in the face of the monkey, or he is puzzled by the monkey's face. Experiments have shown that the development of the ability to decode a human face seems to deprive the ability to decode a monkey's face. Let's conjecture that if we are paying attention to something, then it means we realize that there is something there we do not know or that we may learn something from it.

What is it that attracts our attention? It seems to be something that is new, strange, unusual, surprising, mysterious, unexpected, or odd, at least to an extent. The OPERA's over-the-speed-limit news traveled around the world instantly. Journalists know too well what kinds of news will get their readers' attention—what you think you know could be all wrong! In comparison, we may take things for granted if they appear to be old, common, plain, usual, or dull. Our attention may be torn along other dimensions such as between simplicity and complexity, which is a topic we will return shortly.

1.1.2   Negative Information Attracts More Attention

Unlike no news is good news, it is widely known that any publicity is good publicity. An interesting study published in *Marketing Science* in 2010 further investigated whether it is indeed the case with negative publicity. Jonah Berger, Alan T. Sorensen, and Scott J. Rasmussen revealed a deeper insight—not only whether but also when. In a nutshell, it depends.

They studied weekly national sales of 244 fictions reviewed by the *New York Times*. They found, not surprisingly, that positive reviews significantly boosted sales regardless of the status of the author. In contrast, the effect of negative reviews of a book differs between authors of different status. Negative reviews hurt established authors more by a 15% drop in sales. However, negative reviews boost sales for unknown authors by a 45% increase in sales. More fundamentally, what was really changed by negative reviews was the awareness of unknown authors' fictions. As it appears, to a newcomer, any publicity is indeed good publicity.

Negativity bias is a well-documented phenomenon. Negative information tends to have a greater impact on our cognition and behavior. Controversies, scandals, and mysteries tend to attract a lot of attention. Bad news seems to attract our attention more than good news does. Are we particularly tuned to news of dramatic negative events?

Baumeister, Bratslavsky, Finkenauer, and Vohs (2001) provided a comprehensive review on this topic. One interpretation is that our brains are hardwired to react to negative information after the long evolution in a hunter-gatherer environment. To survive, humans have to attend to anything novel or dramatic immediately.

In an article published in 2001, psychologists Paul Rozin and Edward B. Royzman of the University of Pennsylvania review a wide variety of studies in the literature and developed a taxonomy that can help us better understand the complexity of the phenomenon of negative bias. Rozin and Royzman quoted several widely known expressions in everyday life that underline how we are affected by negative bias. For example, "no news is good news," "a spoonful of tar can spoil a barrel of honey, but a spoonful of honey does nothing for a barrel of tar," "a chain is as strong as its weakest link," and "all happy families resemble one another; every unhappy family is miserable in its own way."

Rozin and Royzman drew our attention to the profound work of Guido Peeters and his colleges, who specifically address negative bias as part of the positive–negative asymmetry in terms of the tension between frequency and importance. Our perceptual and cognitive skills have evolved for it to become effortless to respond to high-frequency low-importance events. In contrast, rare and high-importance events tend to dominate our concerns. Negative events are much rarer than positive ones. Adaptively speaking, the most common events are assumed, but the dangerous and negative events need to be watched closely. Thus, negative bias is rooted in the tension between the more urgent and important events and the less important and common events. In addition, negative events, if left unattended, seem to deteriorate quickly, whereas positive events, without much extra efforts, are unlikely to get even better.

The authors proposed a taxonomy of negative bias phenomena. The taxonomy identifies four aspects of negative bias: negative potency, steeper negative gradients, negative dominance, and negative differentiation. Negative potency refers to the tendency that we are easily disturbed by negative information. The steeper negative gradients indicate that the rate of going from bad to the worst is faster than the rate of making any improvement. Negative dominance specifies one of the Gestalt principles, which we will also discuss shortly. It is not only that the whole is greater than the sum of its parts, but also, if the whole contains positive and negative parts, then holistically, the whole is likely to be perceived negatively altogether. There are numerous examples of how a part can ruin the whole. Negative differentiation underlines the fact that we are much more adept to articulate things that are negative with a finer granularity of detail and a considerably richer vocabulary.

Each of the four aspects can be interpreted in light of the adaptive value of negative bias. For example, negative potency may help us avoid irreversible consequences such as death and fatal accidents. Ways in which we could possibly deal with positive events are generally straightforward. We can approach them and engage them. Should we decide to walk away from a positive event, nothing seems to be there to stop us from doing so. However, it is much more complex when it comes to dealing with a negative situation. We could approach it, but must do so carefully. Sometimes we could freeze and do nothing before we come up with what we should or can do. We may withdraw and flee away from a dangerous situation.

Rozin and Royzman searched for more fundamental reasons that can explain the positive–negative asymmetry. They suggested that our judgment is often made with implicit reference to an ideal situation or a perfect best-case scenario of what could possibly happen. We are anxious in a negative situation because we are drifting away from the ideal and perfect situation that is deeply embedded in our mind. In a positive situation, which is more often than not, we still have yet to reach the perfect state, so we don't seem to have much to get extremely happy about it because we may be still mindful about the state of perfection.

The *gap* between where we are and where we want to be is a notion that will appear again and again in this book. The amount of information, the level of uncertainty, the strength of interest, and the degree of surprise associated with such gaps, physically, conceptually, financially, or psychologically will provide valuable clues and actionable insights on *how* we may use them along the way to reach an even better situation than our current one.

In 2004, The Pew Research Center for the People and the Press published an international survey of a news audience. Emphasis on negative news is the second major complaint among Americans and the British and the third most listed by respondents in Spain, Canada, and France. One possible explanation suggested is that negative information is deemed more valid or true than positive one. Research has shown that negative stimuli are detected more reliably and demand more attention. Negative information is often diagnostic, which may attract additional attention to it.

In a 2006 study, Maria Elizabeth Grabe and Rasha Kamhawi investigated gender differences in processing broadcast news. They particularly focused on valence—the initial response to stimuli as good or bad and whether it is safe to approach or better to stay away—and arousal, the tendency to approach or avoid in response to negative stimuli. Arousal is typically measured at three levels: low, medium, and high. An individual would ignore a stimulus if the arousal level was low. A high arousal level may lead to a more active response to the stimulus. Studies consistently show that women are more likely to avoid moderately arousing negative stimuli than men. In other words, negative stimuli tend to turn most women away, except perhaps weakly negative stimuli, whereas men would be particularly more interested in negative stimuli.

An evolutionary psychology explanation is that the survival of offspring is more dependent on the survival of the mother than the father. Thus, women are responsible for avoiding potential threats in their environment, whereas men are supposed to approach moderate threats in their environment and protect their offspring.

Positively framed news stories often do not produce the same sense of urgency associated with negative news. Grabe and Kamhawi's (2006) experiments revealed that positive framing of news attracts women, yet negative framing attracts men. Traditionally, male journalists and editors have developed news for male readers, who have a cognitive propensity to favor messages with a negative valence. It is reasonable to conjecture that the long tradition in journalism strengthens negative biases in news coverage. More recent research indicated that a positive spin in an otherwise negative story may draw even more attention, such as a heroic move of a vulnerable individual in a tragic story or social cohesion playing a critical role in coping with a natural disaster. Such positive turns not only attract a bigger female audience, but also retain the male audience.

1.1.3   The Myths of Prehistoric Civilization

Why are we fascinated so much by prehistoric civilizations? After all, whatever happened took place a long time ago. Even with today's knowledge of science and technology, we still find it hard to fathom the level of sophistication that prehistoric civilizations have reached. Who built the pyramids in Egypt? How did they achieve the degree of precision that seem to make us puzzled even more? Do they know something that we don't? Are there alternative explanations? Why are we so obsessed with finding a definite answer? Why do we feel a sense of loss when we know that we may never know what truly happened in the remote past or in the remote distance deep in the universe?

Our curiosity is a fundamental drive of our interest. We pay attention to something that we believe we can learn something from. On the other hand, to understand something new, we may need to know enough about something else. In other words, the perceived value of a new piece of information is determined not only by what we already know, but also by how it may be related to what we know. For each individual, what we know is characterized by our mindset. The relevance of information is judged against our mindset. Labels on food packages would convey much needed information to those mindful about their diet, whereas they would be ignored and go unnoticed by others.

Our judgment is guided by our mindset. It makes an individual's behavior meaningful. It makes it possible for others to anticipate how one may react. A closely related concept is a mental model, which is what we think the world is or at least the immediate environment of our concern. A mental model is an understanding of a system. The understanding may be accurate in some aspects of the system but less so in other aspects. The understanding could be wrong altogether, for example, in the OPERA's over the speed limit of light case, and everything that is built on top of it. The accuracy of a mental model is of particular importance in assessing a critical situation. The mental model will guide our course of action. It will help us to identify where we are in a situation and anticipate how things may evolve over space and time. It will serve as the basis for us to identify the gap between our current situation and the goals we want to achieve.

Our mental model can mislead us. Our perceptual and cognitive system can make us believe what we wish to believe. "Seeing is believing" becomes a questionable statement. In order to understand what a mental model can do to us, let's first consider some principles known as collectively as Gestalt principles. Some of these principles are expected to provoke your thinking about the questions we asked earlier. For example, why are we curious about what really happened several hundred thousand years ago?

1.2   GESTALT PRINCIPLES

We are generally happy and satisfied to see smooth surfaces as opposed to rugged and bumpy ones. Smooth and graceful melodies bring peace to our mind. If something is broken, we may wonder if we can do something about it. Gestalt principles underline the tendency of the mind to recognize and accept smooth, complete, and simple patterns. Sometimes such perceptions may lead to illusions that do not actually present in reality. Since our mental models have preferences, we may misperceive the real situation. For example, we hear a stepwise motion in music when there isn't such one per se. We may think we see a face of a human on Mars when shadows play a trick on us.

1.2.1   Closure and Completeness

Human beings have the tendency to perceive patterns out of a variety of incoming information. Figure 1.1 shows two objects, a broken circle and a perfect circle. The perfect circle is considered complete, whereas the broken circle is not. If we are shown the broken circle with a gap that is small enough, our mind may ignore the gap and recognize it as a circle nonetheless. Sometimes the tendency can be so strong that we firmly believe that something must be missing from a potentially perfect circle.

The tendency to see a closed and complete pattern may explain our behavioral and cognitive preferences. A pattern that is offset from a complete and perfect pattern makes us uneasy and unsettled. Mysteries, for example, attract a lot of our attention because we know that we don't have the full story. The state of our mind is not stable. The mind will not settle down until plausible explanations are found.

We are often fascinated by mysteries and eager to find out "what happens" or "what's going on."

When we see or hear part of something, we'd expect it to be complete in one way or the other. For example, in music, a melody may start from one key and eventually return to the same key, thus giving us a sense of completeness—a closure.

We like the shape of circles. It gives us a sense of perfection. When a circle misses a part, we tend to tolerate the missing part and still perceive it as a representation of a full circle. This principle is called the closure principle (Figure 1.2).

The sense of completion plays an important role in our everyday life. We are more interested in the ending of a story than its beginning. An unfinished story makes us feel uneasy and leaves us unsettled. We look forward to seeing what happens next.

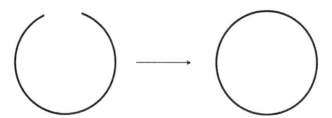

FIGURE 1.1   A preference of the closed circle.

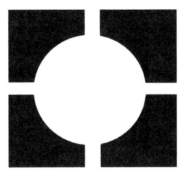

FIGURE 1.2   Are you seeing a circle?

Mysteries attract our attention. We are eager to know what really happened. The hero in the movie Butterfly Effect was able to time travel backward and presumably fix minor problems. However, fixing minor problems messed up a new ending in the future even more. It is challenging to strike the right balance.[1]

When information is simply missing, we may even attempt to fill up the gap with something, anything. In murder cases, wrong persons are identified or falsely identified. In one of the more recent examples, a student of Boston University, Sunil Tripathi, was missing weeks before the Boston Marathon bombing, but in an apparent rush to close the loop, he was wrongly identified as one of the wanted suspects on the popular website Reddit.[2]

On a special section on Reddit called Find Boston Bombers, a crowd-sourced criminal investigation attracted many members who were eager to be the first to identify a suspect. They posted their ideas and, unfortunately, jumped to their conclusions. Sunil Tripathi was not the only one they misidentified. A Blue Rob Man was also misidentified as a suspect by the cyber crowd. His picture even found its way to the front page of the *New York Post*. A few other innocent people were wrongly named in the cyber search. Some of the wrongly identified people received threatening messages.

A common mistake in analyzing a complex case is that we may unknowingly focus on evidence that supports our initial hypothesis but overlook evidence that may challenge our hypothesis. Several elements play an important role in analytic reasoning and decision-making processes, notably, a variety of information, often incomplete; observations, often unreliable; test results, often disputable; and other sources of information and a mental model that can organize, or marshal, the available evidence to a convincing story. There may be gaps from one piece of evidence to the next piece along the reasoning chain. The strength of an argument is as strong as the link that connects the largest gap in the reasoning chain. A particular mental model, or the story, will provide merely one of the potentially many views of the underlying phenomenon. The utility of a mental model is determined by how effectively it serves its purpose. In a courtroom, the question would be how effectively the defense team's

[1] http://www.reelingreviews.com/thebutterflyeffect.htm
[2] http://www.bbc.co.uk/news/technology-22214511

story convinces the members of the jury that it is indeed beyond the reasonable doubt that the defendant did commit the crime as charged. In an experiment such as the OPERA, the investigators need to rule out alternative and more plausible explanations before they make announcements that could upset Einstein.

1.2.2 Continuity and Smoothness

Figure 1.3 illustrates the concept of continuity. On the left, our view is partially blocked by the dark rectangle. Gestalt psychology tells us that we are more likely to believe that it is a perfect circle partially blocked by the rectangle than what is shown on the right—an irregular shape is hidden behind the rectangle. What makes us to think this way?

Simplicity may serve as a more fundamental principle behind the continuity principle. The irregular shape on the right is much more complex than the one on the left. What is the probability of seeing the simpler shape and what is the probability of seeing the more complex shape? Without further information, we simply do not know. In other words, it makes more sense if the chances are 50–50 or evenly distributed. To change this initial belief, we need to take into account additional information. In the next few chapters, we will address this topic from several different perspectives.

Are there alternative ways to explain our preferences toward smooth and graceful shapes over those with more complicated geometry? What if we consider whether all these gestalt principles could be explained by something from a different point of view? Instead of considering shapes and their geometric properties as static and isolated characteristics, let's see if it makes sense to consider various shapes as trails of some invisible objects moving across the space and time? For example, a circle could be the orbit of an object, such as an electron or a planet. An elegant and stylish curve could be the trace of the high-speed movement of a particle or an entire system as a whole. Now reconsidering the continuity principle, is it more likely to see a smooth and continuous track of a moving object or an incomplete and irregular track?

The Gestalt principle on closure is particularly studied in the literature of musical cognition. The closure principle states that the mind may fill in gaps or "missing parts" in order to complete an expected figure. When we listen to a melody, we develop our melodic expectation based on the partial clues in the notes we just heard. In other words, we are building our mental model as the music is being played along.

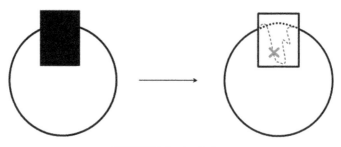

FIGURE 1.3   Continuity.

The structure of music may consist of short-range and long-range components, which provide the template for subsequent variations. Heinrich Schenker (1868–1935), Austrian theorist, was the originator of this type of analysis, known as Schenkerian analysis (Forte & Gilbert, 1982). Schenker's method has revealed that humans prefer smooth progressions over abrupt leaps in music. Our ears sometimes help us fill in the gaps to make the actual music even smoother.

1.2.3   Missing the Obvious

Research has found that we tend to overestimate our perceptual and cognitive abilities. Colin Ware (2008) pointed out that one of the illusions we have is how much we are able to remember from a glance of a complex picture. We remember much fewer details from a picture than we realize. Although we can all agree that one picture is "worth a thousand of words," the fact is that much of what we see will vanish quickly from our perceptual memory and never make it to our memory. We cannot recall details that we did not get in the first place. Then why do we get the feeling that we see everything in the picture?

We can effortlessly direct our attention to any part of a picture, but at any time, the area that actually gets our attention is rather limited. We feel like seeing all the details on the picture at a glance because we could effortlessly attend to any spot on the picture if we want to. There is a significant difference, however, between what we can readily retrieve when needed and what we can still remember after the stimulus is removed.

The accuracy of an eyewitness testimony is a topic of wide interest. How reliable is an eyewitness's testimony? How reliable is our account of what we see in an unanticipated scene? Our memory performs the best when we are neither too stressed nor too relaxed, but right in the middle of the two extremes. Many factors can influence our memory and even distort what we remember. Detectives are always keen on speaking to witnesses before witnesses have a chance to hear different versions of the story from other sources.

Researchers studied the statements from 400 to 500 witnesses who were waiting for the arrival of President Kennedy at Dealey Plaza on November 23, 1963.[3] Researchers were interested in how accurately witnesses described what they saw or heard. How many gunmen were there at Dealey Plaza? How many gunshots were heard? Where were the gunshots originated? The study found that witnesses gave completely different answers to these questions.

The accuracy of eyewitness testimony is also questionable as details are revealed by recorded experiments. In one experiment, a staged assault of a professor took place in front of 141 unsuspecting witnesses. The incident was recorded on videotape. Immediately after the incident, everyone at the scene was asked to describe the incident in detail. Descriptions given were mostly inaccurate when compared to the videotape. Most people overestimated the duration of the assault by two and half times longer. The attacker's weight was overestimated, but his age was underestimated. After seven

[3] http://mcadams.posc.mu.edu/zaid.htm

weeks, the witnesses and the professor himself were asked to identify the attacker from a group of photographs—60% of them picked the wrong guy and 25% mistook an innocent bystander as the attacker.

Researchers have long been interested in what separates the expert from the novice in their ability to recall what they see, and the role of expertise in explaining such differences. Chase and Simon (1973) conducted the best-known experiment concerning expert and novice chess players.[4] In their study, a chess master was regarded as an expert player. Between the chess master and a novice player, there was an intermediate-level player who was better than a novice player but not as good as a chess master. First, chess pieces were positioned according to a real middle game. All the players were given five seconds to view the positions. Then the chessboard was covered. Players were instructed to reconstruct the positions of these pieces on a different chessboard based on their memory. The performance was assessed in terms of the accuracy of their reconstruction. The master player performed the best, followed by the intermediate-level player. The master player correctly placed twice as many pieces as the intermediate-level player, who in turn correctly placed twice as many pieces as the novice player.

The second part of Chase and Simon's chess experiment was the same as the first one except this time the positions of chess pieces were arranged randomly rather than according to real chess games. This time, all players performed with the same level of accuracy—the number of correctly recalled positions was much less than what the expert players could recall from settings based on real games.

Generally speaking, experts always demonstrate superior recall performance within their domain of expertise. As the chess experiments show, however, the superior performance of experts is unlikely to do with whether they have a better ability to memorize more details. Otherwise, they would perform about the same regardless of whether the settings were random or realistic. So what could separate an expert from a novice?

Researchers have noticed that experts organize their knowledge in a much more complex structure than novices do. In particular, a key idea is chunking, which organizes knowledge or other types of information at different levels of abstraction. For example, geographic knowledge of the world can be organized into levels of country, state, and city. At the country level, we do not need to address details regarding states and cities. At the state level, we do not need to address details regarding cities. In this way, we find it easier to handle the geographic knowledge as a whole. A study of memory skills in 1988 analyzed a waiter who could memorize up to 20 dinner orders at a time (Ericsson & Polson, 1988). The waiter used mnemonic strategies to encode items from different categories such as meat temperature and salad dressing. For each order, he encoded an item from a category along with previous items from the same category. If an order included blue cheese dressing, he would encode the initial letter of the salad dressing along with the initial letters from previous orders. So orders of blue cheese, oil vinegar, and oil vinegar would become a BOO. The retrieval mechanism would be reflected at encoding and retrieval. Cicero (106–43 BC), one of the greatest

---

[4] http://matt.colorado.edu/teaching/highcog/fall8/cs73.pdf

orators in ancient Rome, was known for his good memory. He used a strategy called memory palaces to organize information for later retrieval. One may associate an item with a room in a real palace, but in general any structure would serve the purpose equally well. The chunking strategy is effectively the same as using a memory palace.

1.3 AESTHETICS

What is it that gives us the sense of elegance, beauty, or sublime? What makes a melody pleasant to the ear? What makes a painting pleasant to the eye? What makes a human face look attractive?

What makes a natural environmental scene breathtaking? We do not expect to find a single answer for all these questions, but the literature from a diverse range of disciplines seems to point to some consistent patterns.

In a 2009 study, Martina Jakesch and Helmut Leder found that uncertainty in the form of ambiguity is playing a major role in aesthetic experiences of modern art. When they tested whether liking and interestingness of a picture were affected by the amount of information, the amount of matching information, or levels of ambiguity defined as the proportion of matching to nonmatching statements apparent in a picture, the ambiguity was the only independent variable that predicted differences in liking and interestingness. They concluded that ambiguity is an important determinant of aesthetic appreciation and a certain level of ambiguity is appreciable.

1.3.1 The Golden Ratio

The golden ratio is widely believed to hold the key to the answers to at least some of the questions provided earlier. But if it is indeed the case, what makes the golden ratio so special, and why are human beings so responsive to its presence?

Figure 1.4 illustrates how the golden ratio is calculated. Consider a unit length string, that is, the length of string is 1, but the unit does not matter (it could be 1 meter, 1 mile, or even 1 light year). There is a magic point on the string that divides the string into two

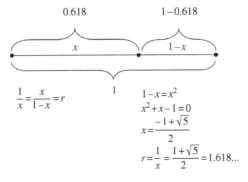

FIGURE 1.4   The golden ratio calculation.

parts, creating a division that is considered perfect. Suppose one part of the division has the length x, and it follows that the other part's length must be $1-x$. The value of the golden ratio is given by the equation $(1/x) = (x/(1-x))$. Solving the equation for x, the golden ratio is $1/x$, which is approximately 1.618.

In 2012, a British company, Lorraine Cosmetics, sponsored a contest to find Britain's perfect face.[5] Contestants were required to have completely natural faces, no makeup, no Botox, and no cosmetic surgery. The contest attracted 8045 contestants. Three finalists were selected from the 18 to 40 age group and three from the 41 and over group by a panel. The six finalists were voted by the public. An 18-year-old student, Florence Colgate, from Kent, England, was the winner.

A video demonstrated that there are as many as 24 ways to fit rectangles of a golden ratio division on her facial features such as eyes, pupils, lips, and so on. The British contest provides a concrete example of how the golden ratio is connected to the objective aspect of what the beauty is.

1.3.2 Simplicity

The word *elegance* is defined in dictionaries as the quality of being pleasingly ingenious, simple, and neat. An elegant move refers to the graceful and smoothness of a movement. Complexity, on the other hand, is not usually considered an appealing property. Not only do we prefer simplicity, but we also seek ways that may reduce the complexity of an entity or a system. Any suggestions that may increase the current complexity tend to be carefully justified or penalized.

Regularity is often considered to be an aesthetic characteristic. It frequently serves as a complexity reduction mechanism. As the Britain's perfect face contest revealed, one may repeatedly apply the golden ratio rule to configure what the most beautiful face should look like. No wonder it was sponsored by a cosmetics company.

The presence of regularity opens the door to improve our understanding of something complex through an algorithmic perspective. Algorithms are recipes that prescribe a sequence of actions. Completing these actions will accomplish a predefined goal. Algorithmic thinking does not necessarily require a computer. It can be a rigorous logical reasoning process that explains how things work. In the next few chapters of the book, we will encounter other examples of how algorithmic thinking provides the key to the understanding of a complex system.

Jürgen Schmidhuber is a computer scientist and artist. He developed a theory to explain what makes something interesting, especially in low-complexity art. According to his theory, our level of interest is affected by what we already know and what we see. It is a function of the gap between our prior knowledge and the new information. If the gap is very small, the new information is likely to be seen as trivial. If the gap is very large, the relevance of the new information becomes questionable. Thus, the region of interest is likely to fall between the two extremes. So what makes the new information interesting if the size of the gap is just right?

[5] http://www.goldennumber.net/facial-beauty-golden-ratio-florence-colgate/

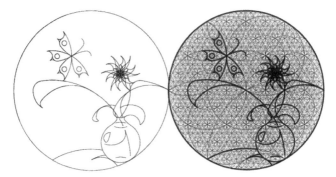

FIGURE 1.5 The drawing on the left consists of segments from various sized circles. Source: Figure 2a, 2b in Schmidhuber (1997). http://www.idsia.ch/~juergen/beauty.html. Reproduced with permission. © Schmidhuber 1994.

Schmidhuber's theory explains why the gap is an important factor. He suggests that in such a situation, one may find the new information interesting if it appears that there would be a new way to consolidate the new information with our existing knowledge without increasing the overall complexity. The process of consolidating the information is a learning process. It is a process that encodes the information so that it fits to our knowledge, and the encoding needs to be a simpler way to reconstruct what we know.

Using a compression algorithm as an analog, Schmidhuber explains the nature of the encoding process. We find something interesting because we sense that we may be able to find a more efficient way to compress the new information with our prior knowledge. If we see no sign of the potential to improve the current compression, then we are less likely to find the new information interesting. In contrast, if the compression rate can be indeed improved substantially, then the tension generated by the gap characterizes the degree of interest. In Schmidhuber's own words, interest-ingness corresponds to the first derivative of a perceived gap that can be bridged by learning, that is, finding a more efficient encoding algorithm (Figure 1.5).

In terms of mental models, Schmidhuber's theory is really saying that we find new information interesting if there is a seemingly bridgeable gap between our current mental model and the new information. Studies of radical ideas in technological inventions echo the wisdom. Good ideas may be drawn from disparate patches of knowledge, but these patches should not be too close or too farther away from each other either. What is the level of interest generated by the OPERA over the speed limit of light among physicists and the general public? To physicists, the level of interest was not particularly high; many stressed the need to validate the initial finding carefully in multiple ways. The gap between the news and the current knowledge of physics seems to be too large to be interesting because there is no known mechanism in physics that would explain what the OPERA claimed. To the general public, the perceived gap is perhaps somewhat narrower because the understanding is at a higher level of granularity. Although no one can explain how it was possible to travel faster than the speed of light, that's something for physicists to worry about, and it sounds amusing that Einstein could be wrong after all these years).

The encoding theory, on the other hand, is not elaborated enough to explain substantial changes in our beliefs, for example, the type of scientific revolutions described by Thomas Kuhn, a philosopher of science. If the OPERA's measurement was not flawed, then physicists would become less comfortable, and they would have to reconcile the contradictory observation and the long-standing theory. If that was the case, then physicists may find themselves considering an alternative theory to Einstein's relativity theory. The level of interest then would be very high and justifiably high.

1.3.3   Regularity

Finding a simpler encoding is easier if regularity is evident. Regularity signifies that it is indeed possible to reduce the overall complexity substantially. The relationship between encoding algorithms of different efficiency is similar to that between different arithmetic operations such as addition and multiplication. Adding the number 10 for five times can be expressed as $10 + 10 + 10 + 10 + 10$. Alternatively, it can be expressed more concisely in terms of a multiplication of $10*5$. The multiplication is a more efficient compression method than the addition method. It is possible because of its regularity.

Similarly, compression algorithms for images and other types of computer files can be seen as a complexity reduction process. The same strategy has been used in reconstructing a dataset on the fly. Instead of storing the actual data directly, an alternative is to store the algorithm that can generate the dataset when the data is needed. The size of the data-reconstruction algorithm is usually much smaller than the resultant dataset it is designed to generate.

Regularity therefore plays a key role in whether the new information is perceived to be interesting or not. A series of completely random numbers has no regularity. It is impossible to devise an algorithm that would be able to reproduce these random numbers without retaining as much data as the original dataset. In contrast, if we know that a series of numbers is in fact a Fibonacci series, then all we need to remember is how these numbers can be generated from the previous ones. The first two Fibonacci numbers are 0 and 1, followed by numbers such that each number is the sum of the previous two numbers. The third Fibonacci number is 1, which is the sum of 0 and 1. The fourth number is 2, which is the sum of 1 and 1. This type of regularity is commonly seen in mathematics, physics, and several other disciplines.

Ratios between adjacent Fibonacci numbers approximate to the golden ratio $\text{phi} = \left(1 + \sqrt{5}\right) / 2 = 1.6180339887...$ (Table 1.1).

Regularity implies simplicity and elegance. If we see regular patterns in a phenomenon and even though we may not immediately understand what they are, we would find it interesting because the regularities mean that the complexity can be reduced by obtaining an understanding of the new regularity.

Regularities may not be always obvious from a particular perspective. Sometimes the same phenomenon may appear much simpler in one perspective than another.

Table 1.1   The ratios between adjacent
Fibonacci numbers approaching the golden ratio

F_n	F_{n+1}	$\dfrac{F_{n+1}}{F_n}$
0	1	
1	1	1.00000
1	2	2.00000
2	3	1.50000
3	5	1.66667
5	8	1.60000
8	13	1.62500
13	21	1.61538
21	34	1.61905
34	55	1.61765
55	89	1.61818
89	144	1.61798
144	233	1.61806
233	377	1.61803
377	610	1.61804
610	987	1.61803
987	1597	1.61803

Somehard questions may suddenly become obvious to answer purely because a more appropriate viewpoint is chosen.

For example, the spread of infectious diseases such as influenza has shown to be quite random on the world geographic map. The patterns of the spread become much simpler if the airline travel distance is used instead of geographic distance—the alternative perspective reveals the spread as a continuously enlarged circle centered at the origin of the flu.

1.3.4   Beauty

Human faces have been the center of studies of what the beauty means. The most famous study of human faces can be traced back to the pioneering investigation by Francis Galton in late 1870s. The original idea was to see whether it would be feasible to identify any common characteristics from those who committed crimes. In today's terms, the goal was similar to that of profiling people who are more likely to commit crimes than the rest of the population.

Francis Galton studied a variety of information about a large number of criminals, including their photographs, which was provided to him by Sir Edmund DuCane. Both of them were convinced that if criminals share some common features, then such features could be valuable to identify the criminals-to-be earlier on.

Galton was particularly interested in whether criminals would have any facial features in common. In order to make common features stand out, he devised a

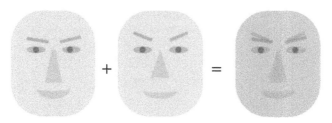

FIGURE 1.6   Photographs are combined to identify common features.

"Composite Portraiture" method to overlay multiple photographs of individuals' faces one by one (Galton, 1879). Photographs of different criminals were combined by controlling the time of exposure to produce a single blended image. He expected that common facial features would be repeatedly reinforced so that they would appear more distinct than less common features, which would appear blurred due to the diversity (Figure 1.6). Details of his experiments and much more information about him are available at http://galton.org/.

He was surprised by what he found—the more photos were added together in the process, the better looking the combined face would get. This observation prompted him to propose that the beauty probably has something do with the "average" of facial features across the population.

Galton's idea was intriguing. It suggested a potentially simple and elegant way to interpret what the beauty is. More importantly, his method provides a method that could bridge the gap between a hypothesis and a large-scale test.

Large-scale tests of his hypothesis didn't come immediately. His idea was well ahead of his time. He didn't have the necessary technology to scale up his manual procedure of photographically combining individual photographs. It wasn't as easy to collect standard portrait photos of a large number of individuals as it is today, either.

Galton also tried to apply his Composite Portraiture method to identify the common profile of people who had chronic illnesses. The results were inconclusive largely due to the small sample size. Modern studies have shown that with very large samples, there is indeed a weak but significant link between physical appearances and traits like criminality.

Galton's idea attracted numerous studies. It remains to be a popular topic today, if not more popular. Significant improvements have been made in pursuit of the idea with increasingly larger sample sizes and more sophisticated calibration techniques. In 2013, the number of faces used to generate the synthesized face exceeded 5000 individuals.

On the other hand, the idea that the beauty represents something fundamental to the entire population has been criticized from various perspectives. Suppose we assemble a representative face from the better looking half of the population and compare it with a representative face of the remaining half of the population. We would expect to notice some differences. However, according to the literature, neither infants nor adults could tell much difference.

In a classic study of facial attractiveness, Perrett, May, and Yoshikawa (1994) tested the Averageness Hypothesis—attractive faces are the average face—in order to establish if averageness really is the critical determinant of the attractiveness of faces. Their findings rejected the Averageness Hypothesis as they found that highly attractive faces deviate systematically from the average.

Earlier research suggested that a smiling face seems to be more attractive. Does it make a difference if the smiling person is happy in general or the person is particularly happy to see you and smiling at you? Benedict Jones, DeBruine, Little, Conway, and Feinberg in a 2006 study demonstrated that attraction is influenced by how much an individual appears interested in interacting with you. In other words, attraction is not solely determined by physical beauty. So the beauty is personal after all.

A different approach to study aesthetics is to identify patterns computationally from physical features of images. Datta, Joshi, Li, and Wang (2006) conducted a quantitative study of aesthetics in photographic images with an array of physical features. They used a variety of image processing techniques to divide a photo into smaller patches, to measure the brightness, to calculate the golden ratio of the saturation, and to compute about 40 other attributes from an image. Note that the photos were not limited to human faces. Statistically, several computational features appear to play a major role in association with attractive ratings of an image, notably the average saturation for the largest patch (feature 31), the average brightness (feature 1), the low depth of field indicator for saturation (feature 54), the average hues for the third largest patch (feature 28), and the relative size of the third largest patch with respect to the size of the entire image (feature 43). The effect of whether the focal point of the image divides the entire image by the golden ratio was found further down the list, as the 15th strongest effect. Many of the influential variables appear to do with the level of saturation of prominent components of an image.

Natural selection and the survival of the fittest are widely known. In the context of evolutionary genetics, the fitness of an individual in a population is his/her potential to breed their offspring. It is natural to attempt to explain beauty in terms of fitness, that is, the beauty is the effect of the best genetic combination. One can envisage a variety of experiments along with this line of reasoning. In a recent example, twins with different smoking history were compared to see if they were visibly different. The photos of their faces appear to show that smoking and the speed of aging may be connected. The fitness is a profound topic that plays a central role in the following chapters. We will return to this intriguing topic in much more detail.

By the way, Charles Darwin was Francis Galton's cousin.

1.4 THE INDEX OF THE INTERESTING

The late sociologist Murray S. Davis developed some intriguing insights into why we are interested in some (sociological) theories but not others (Davis, 1971). In a nutshell, we pay attention to a theory not really because the theory is true and valid; instead, a theory is getting our attention because it may change our beliefs.

1.4.1 Belief Updates

Although his work focused on sociological theories, the insights are broad-ranging. According to Davis, "the truth of a theory has very little to do with its impact, for a theory can continue to be found interesting even though its truth is disputed—even refuted!" A theory is interesting to the audience because it denies some of these assumptions or beliefs to an extent. If a theory goes beyond certain points, it has gone too far, and the audience will lose their interest.

The interestingness is about what draws our attention. Davis asked: where was the attention before it was engaged by the interesting? Before the state of attention, most people are not attentive to anything in particular, and we take many things for granted. Harold Garfinkel (1967) called this state of low attention "the routinized taken-for-granted world of everyday life." If some audience finds something interesting, it must have stood out in their attention in contrast to the taken-for-granted world—it stands out in their attention in contrast to propositions that a group of people have taken for granted. The bottom line is that a new theory will be noticed only when it denies something that people take for granted, such as a proposition, an assumption, or a commonsense.

One of the keynote speakers at IEEE VisWeek 2009 at Atlantic City told the audience how to tell an engaging story. The template he gave goes like this: here comes the hero; the hero wants something, but the hero is prevented from getting what he wants; the hero wants it badly; and finally, the hero figures out an unusual solution. The storytelling rubric has something in common with why a theory is interesting. Both are connected to our curiosity of something unexpected.

Davis specifies a rhetorical routine to describe an interesting theory: (1) The author summarizes the taken-for-granted assumptions of his audience. (2) He challenges one or more conventional propositions. (3) He presents a systematically prepared case to prove that the wisdom of the audience is wrong. He also presents new and better propositions to replace the old ones. (4) Finally, he suggests the practical consequences of new propositions.

David realized that the nature of interesting can be explained in terms of a dialectical relation between what appears to be and what could really exist. In his terminology, the appearances of a phenomenon experienced through our senses are phenomenological, whereas intrinsic characteristics of the phenomenon are ontological. An interesting theory, therefore, is to convince the audience that the real nature of a phenomenon is somewhat different from what it seems to be. He identified 12 logical categories that capture such dialectical relations (Table 1.2).

1.4.2 Proteus Phenomenon

Proteus is a sea god in Greek Mythology. He could change his shape at will. The Proteus phenomenon refers to early extreme contradictory estimates in published research. Controversial results can be attractive to investigators and editors. Ioannidis and Trikalinos (2005) tested an interesting hypothesis that the most extreme opposite results would appear very early as data accumulate rather than late. They used meta-analyses of studies on genetic associations from MEDLINE and meta-analyses of randomized trials

Table 1.2   Dialectical relations between taken-for-granted assumptions and new propositions

Phenomenon	Dialectical relations			
Single	Organization	Structured	⟵⟶	Unstructured
	Composition	Atomic	⟵⟶	Composite
	Abstraction	Individual	⟵⟶	Holistic
	Generalization	Local	⟵⟶	General
	Stabilization	Stable	⟵⟶	Unstable
	Function	Effective	⟵⟶	Ineffective
	Evaluation	Good	⟵⟶	Bad
Multiple	Co-relation	Interdependent	⟵⟶	Independent
	Coexistence	Coexist	⟵⟶	Not coexist
	Covariation	Positive	⟵⟶	Negative
	Opposition	Similar	⟵⟶	Opposite
	Causation	Independent	⟵⟶	Dependent

of health-care interventions from the Cochrane Library. They evaluated how the between-study variance for studies on the same question changed over time and at what point the studies with the most extreme results had been published. The results show that for genetic association studies, the maximum between-study variance was more likely to be found early in the 44 meta-analyses and 37 in the health-care intervention case. The between-study variance decreased over time in the genetic association studies, which was statistically significant.

1.4.3   Surprises

Information entropy is a useful system-level metric of fluctuations of the overall information uncertainty in a large-scale dynamic system. Figure 1.7 shows how the information entropy of terrorism research changes over 18 years based on keywords assigned to scientific papers on the subject between 1990 and the first half of 2007. Entropies are computed retrospectively based on the accumulated vocabulary throughout the entire period. Two consecutive and steep increases of entropy are prominently revealed, corresponding to 1995–1997 and 2001–2003. The eminent increases of uncertainty send a strong message that the overall landscape of terrorism research must have been fundamentally altered. The unique advantage of the information-theoretic insight is that it identifies emergent macroscopic properties without overwhelming analysts with a large amount of microscopic details. Using the terminology of information foraging, these two periods have transmitted the strongest information scent. Note that using the numbers of unique keywords fails to detect the first period identified by information entropy. Subsequent analysis at microscopic levels reveals that the two periods are associated with the Oklahoma City bombing in 1995 and the September 11 terrorist attacks in 2001.

Information indices allow us to compare the similarity between different years. Figure 1.8 shows a 3D surface of the K–L divergences between distributions in

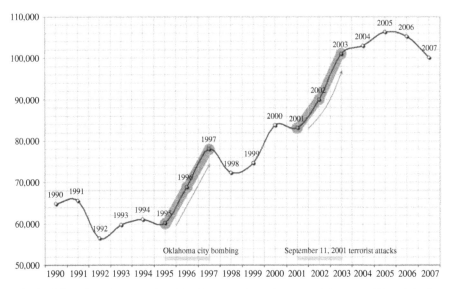

FIGURE 1.7 Information entropies of the literature of terrorism research between 1990 and the first half of 2007. The two steep increases correspond to the Oklahoma City bombing in 1995 and the September 11 terrorist attacks in 2001. Source: From Chen (2008).

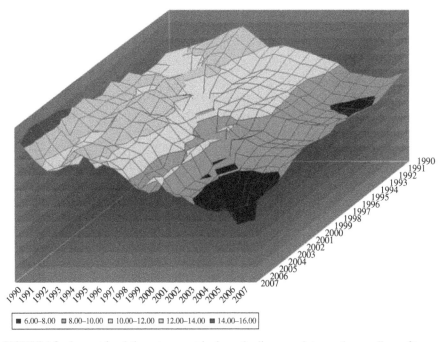

FIGURE 1.8 Symmetric relative entropy matrix shows the divergence between the overall use of terms across different years. The recent few years are most similar to each other. The boundaries between areas in different colors indicate significant changes of underlying topics. Source: From Chen (2008).

different years. The higher the elevation is, the more difference there is between two years of research. For example, the blue area has the lowest elevation, which means that research is more similar in the recent three years than earlier years. Three-dimensional representations of this kind share a common metaphor of a landscape, which forms valleys and peaks. The intuitive metaphor suggests to the viewer that operations such as moving across the landscape, moving away from a valley, or climbing up a hill may become meaningful in the context defined by such landscape models. In this example, the darkest area is a valley, where research is relatively stable in adjacent years. In contrast, the areas in lighter colors are peaks and ranges, where research is experiencing more active changes. Fitness landscapes are a core topic in our book. We will explore a variety of studies using the metaphor of fitness landscapes in later chapters.

Information-theoretic techniques provide not only a means of addressing macroscopic questions, but also a way to decompose and analyze questions at lower levels of aggregation. Given that we have learned that there are two distinct periods of fundamental transformation in terrorism research, the next step is to understand what these changes are in terms of their saliency and novelty. Different distributions may lead to the same level of entropy. In order to compare and differentiate different distributions, one can use information-theoretic metrics such as information bias, which measures the degree to which a subsample differs from the entire sample that the subsample belongs to. High-profile thematic patterns can be easily identified in terms of term frequencies. Low-profile thematic patterns are information-theoretic outliers from the mainstream keyword distributions. Low-profile patterns are equally important as high-profile patterns in analytical reasoning because they tell us something that we are not familiar with and something novel.

Informational bias $T(a{:}B)$ is defined as follows, where a is a subsample of the entire sample B. p_{ab}, p_a, p_b, and $p_{b|a}$ are corresponding probabilities and conditional probabilities. $H_a(B)$ is the conditional entropy of B in the subsample. We take the distribution of a given keyword and compare it to the entire space of keyword distributions:

$$T(a:B) = \frac{1}{p_a} \sum_b p_{ab} \log_2 \frac{p_{ab}}{p_a p_b}$$

$$T(a:B) = \sum_b p_{b|a} \log_2 \frac{p_{b|a}}{p_b} = -\sum_b p_{b|a} \log_2 p_b - H_a(B)$$

One may study both high- and low-profile patterns side by side. The network shown in Figure 1.9 consists of keywords that appeared in articles published in 1995, 1996, and 1997, corresponding to the first period of substantial change in terrorism research. High-profile patterns are labeled in black, whereas low-profile patterns are labeled in italic. High-profile patterns help us to understand the most salient topics in terrorism research in this period of time. For example, *terrorism, posttraumatic stress disorder, terrorist bombings,* and *blast overpressure* are the most salient ones. The latter two are closely related to the Oklahoma City bombing event, whereas *posttraumatic stress disorder* is not directly connected at this level. In contrast, low-profile patterns include *avoidance symptoms, early intrusion,* and *neuropathology*. These terms are unique with

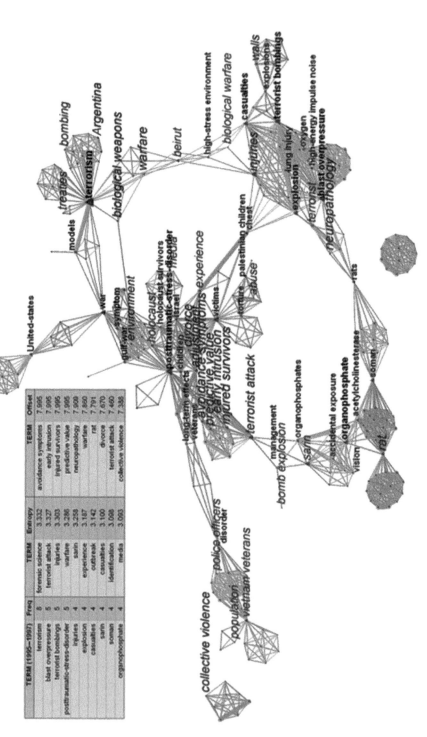

FIGURE 1.9 A network of keywords in the terrorism research literature (1995–1997). High-frequency terms are shown in boldface, whereas outlier terms identified by informational bias are shown in italic. Source: From Chen (2008).

reference to other keywords. Once these patterns are identified, analysts can investigate even further and make informed decisions. For example, one may examine whether this is the first appearance of an unexpected topic or whether the emergence of a new layer of uncertainty to the system at this point makes perfect sense.

1.4.4 Connecting the Dots

The investigation of 9/11 terrorist attacks has raised questions on whether the intelligence agencies could have connected the dots and prevented the terrorist attacks (Anderson, Schum, & Twining, 2005). Prior to the 9/11 terrorist attacks, several foreign nationals enrolled in different civilian flying schools to learn how to fly large commercial aircraft. They were interested in learning how to navigate civilian airlines, but not in landings or takeoffs. And they all paid cash for their lessons. What is needed for someone to connect these seemingly isolated dots and reveal the hidden story?

In an intriguing *The New Yorker* article, Gladwell differentiated puzzles and mysteries with the stories of the collapse of Enron (Gladwell, 2007). To solve the puzzle, more specific information is needed. To solve a mystery, one needs to ask the right question. Connecting the dots is more of a mystery than a puzzle. Solving mysteries is one of the many challenges for visual analytic reasoning. We may have all the necessary information in front of us and yet fail to see the connection or recognize an emergent pattern. Asking the right question is critical to stay on track.

In many types of investigations, seeking answers is only part of the game. It is essential to augment the ability of analysts and decision makers to analyze and assimilate complex situations and reach informed decisions. We consider a generic framework for visual analytics based on information theory and related analytic strategies and techniques. The potential of this framework to facilitate analytical reasoning is illustrated through several examples from this consistent perspective.

In information theory, the value of information carried by a message is the difference of information entropy *before* and *after* the receipt of the message. Information entropy is a macroscopic measure of uncertainty defined on a frequency or probability distribution. A key function of an information-theoretic approach is to quantify discrepancies of the information content of distributions. Information indices, such as the widely known Kullback–Leibler (K–L) divergence (Kullback & Leibler, 1951), are entropy-based measures of discrepancies between distributions (Soofi & Retzer, 2002).

The K–L divergence of probability distribution Q from probability distribution P is defined as follows:

$$\text{Divergence}_{x-z}(P:Q) = \sum_{i} P(i) \log \frac{P(i)}{Q(i)}$$

The divergence measures the loss of information if Q is used instead of P, assuming P is the true distribution. Information entropy can be seen as the divergence from the uniform distribution. This is consistent with the common interpretation of information entropy as a measure of uncertainty or the lack of uniformity.

A useful alternative interpretation of the K–L divergence is the expected extra message length to be entailed in communication if the message is transmitted without using a coding scheme based on the true distribution. In computer science, the Kolmogorov complexity of an object, also known as algorithmic entropy or program-size complexity, measures the amount of computational resources required to specify the object.

A salient feature or pattern is prominent in the sense that it stands out perceptually, conceptually, and/or semantically. In contrast, novelty characterizes the uniqueness of information. A landmark has a high saliency in its skyline, whereas the novelty of a design is how unique it is in comparison to others. It is often effortless for humans to spot salient or novel features visually. However, it is a real challenge to identify these features computationally because these are emergent macroscopic features in nature as opposed to specific microscopic ones.

From an information-theoretic point of view, saliency can be defined as statistical outliers in a semantic and/or visual feature space. Novelty, on the other hand, can be defined as statistical outliers along a specific dimension of the space, such as the temporal dimension.

Itti and Baldi developed a computational model to detect surprising events in video (Itti & Baldi, 2005). Their goal was to find surprising scenes. They focused on how one's belief changed between two distinct frames. One's belief is transformed from a prior distribution P(Model) to a posterior distribution P(Model | Data). The difference between prior and posterior distributions over all models is measured by relative entropy, that is, the K–L divergence. Surprise is defined as the average of the log-odds ratio with respect to the prior distribution over the set of models. The higher the K–L scores, the more discriminate the detection measures are. Surprises identified by the computational model turned out to be a good match to human viewers' eye movements on video images.

1.5 SUMMARY

We introduced a series of questions from a diverse spectrum of perspectives to underline what might be the role of information behind so many phenomena that we often take for granted. At the first glance, questions concerning what attract our attention and what we perceive as beauty may be philosophical and subjective. Our review of various studies has revealed that how these questions have been addressed may be insightful for us to better understand the value of information and how it interacts with our knowledge.

The explanation that we are attracted by simplicity and regularity is among the most compelling ones. The implications of regularity are particularly interesting from a computational and constructivist perspective, which not only explains the underlying mechanisms but also demonstrates how we could actually reach the goal step by step.

The puzzle of why we seem to be obsessed with the so-called negative bias leads us to explore further for possible sources where it might come from. It is insightful to learn about the gender difference in this context, that is, males are more likely than females to be tuned to the negative signals. This understanding takes us even further, and an evolutionary psychological explanation appears to make sense. Whether one can scientifically verify these evolutionary explanations aside, in the next chapter, we will look at possible sources of regularity and other pleasing patterns and what they mean to us in terms of what we can do or act upon them.

A recurring topic in this chapter is the notion of a gap. The gap can be physical, conceptual, or informational. The potential of how we may handle a perceived or otherwise sensed gap attracts our attention. Another concept that seems always lurking with these topics is the notion of a mental model. To identify a gap or measure a change, one needs to have a reference point. The concepts of mental models and gaps will serve as some of the most fundamental building blocks for the rest of the book.

BIBLIOGRAPHY

Aday, S. (2010). Chasing the bad news: An analysis of 2005 Iraq and Afghanistan war coverage on NBC and Fox News Channel. *Journal of Communication, 60*(1), 144–164.

Anderson, T., Schum, D., & Twining, W. (2005). *Analysis of evidence* (2nd ed.). Cambridge, England: Cambridge University Press.

Baumeister, R. F., Bratslavsky, E., Finkenauer, C., & Vohs, K. D. (2001). Bad is stronger than good. *Review of General Psychology, 5*, 323–370.

Berger, J., Sorensen, A. T., & Rasmussen, S. J. (2010). Positive effects of negative publicity: When negative reviews increase sales. *Marketing Science, 29*(5), 815–827.

Carlson, A. (1979). Appreciation and the natural environment. *Journal of Aesthetics and Art Criticism, 37*(3), 267–275.

Chase, W. G., & Simon, H. A. (1973). Perception in chess. *Cognitive Psychology, 4*, 55–81.

Chen, C. (2008). An information-theoretic view of visual analytics. *IEEE Computer Graphics & Applications, 28*(1), 18–23.

Datta, R., Joshi, D., Li, J., & Wang, J. Z. (2006). Studying aesthetics in photographic images using a computational approach. *Computer Vision – ECCV 2006*. Lecture Notes in Computer Science, Vol. *3953*, 288–301.

Davis, M. S. (1971). That's interesting! Towards a phenomenology of sociology and a sociology of phenomenology. *Philosophy of the Social Sciences, 1*(2), 309–344.

Ericsson, K. A., & Polson, P. G. (1988). An experimental analysis of the mechanisms of a memory skill. *Journal of Experimental Psychology: Learning, Memory, and Cognition, 14*, 305–316.

Forte, A., & Gilbert, S. E. (1982). *Introduction to Schenkerian analysis*. New York: W. W. Norton & Company.

Galton, F. (1879). Composite portraits made by combining those of many different persons into a single figure. *Journal of the Anthropological Institute, 8*, 132–148.

Garfinkel, H. (1967). *Studies in ethnomethodology*. Englewood Cliffs, NJ: Prentice Hall.

Gladwell, M. (2007, January 8). Open secrets: Enron, intelligence, and the perils of too much information. *The New Yorker*. Retrieved March 1, 2007, from http://www.newyorker.com/reporting/2007/01/08/070108fa_fact

Grabe, M. E., & Kamhawi, R. (2006). Hard wired for negative news? Gender differences in processing broadcast news. *Communication Research, 33*(5), 346–369.

Graham, D. J., & Redies, C. (2010). Statistical regularities in art: Relations with visual coding and perception. *Vision Research, 50*, 1503–1509.

Gregory, W. (2009). System and method for adaptive segmentation and motivic identification. U.S. Patent No. 8,084,677. Retrieved June 1, 2013, from http://www.google.com/patents/WO2009085054A1. Accessed on March 10, 2014.

Ioannidis, J. P., & Trikalinos, T. A. (2005). Early extreme contradictory estimates may appear in published research: The Proteus phenomenon in molecular genetics research and randomized trials. *Journal of Clinical Epidemiology, 58*, 543–549.

Ito, T. A., Larsen, J. T., Smith, N. K., & Cacioppo, J. T. (1998). Negative information weighs more heavily on the brain: The negativity bias in evaluative categorizations. *Journal of Personality and Social Psychology, 75*(4), 887–900.

Itti, L., & Baldi, P. (2005, June). *A principled approach to detecting surprising events in video* (Vol. 1, pp. 631–637). Paper presented at the proceedings of the IEEE conference on Computer Vision and Pattern Recognition (CVPR), June 20–26, 2005, San Diego, CA.

Jakesch, M., & Leder, H. (2009). Finding meaning in art: Preferred levels of ambiguity in art appreciation. *Quarterly Journal of Experimental Psychology, 62*(11), 2105–2112.

Jones, B. C., DeBruine, L. M., Little, A. C., Conway, C. A., & Feinberg, D. R. (2006). Integrating physical gaze direction and expression with physical attractiveness when forming face preferences. *Psychological Science, 17*(7), 588–591.

Kullback, S., & Leibler, R. A. (1951). On information and sufficiency. *Annals of Mathematical Statistics, 22*, 79–86.

Langlois, J. H., Kalakanis, L., Rubenstein, A. J., Larson, A., Hallam, M., & Smoot, M. (2002). Maxims or myths of beauty? A meta-analytic and theoretical review. *Psychological Bulletin, 126,* 390–423. http://faceresearch.org/students/papers/Langlois_2000.pdf. Accessed on March 10, 2014.

Langlois, J. H., & Roggman, L. A. (1990). Attractive faces are only average. *Psychological Science, 1*(2), 115–121.

Little, A. C., & Jones, B. C. (2003). Evidence against perceptual bias views for symmetry preferences in human faces. *Proceedings of the Royal Society London B, 270*, 1759–1763.

Livio, M. (2002). *The golden ratio and aesthetics*. http://plus.maths.org/content/os/issue22/features/golden/index. Accessed on March 10, 2014.

Perrett, D. I., May, K. A., & Yoshikawa, S. (1994). Facial shape and judgements of female attractiveness. *Nature, 368*, 239–242.

Ramachandran, V. S., & Hirstein, W. (1999). The science of art: A neurological theory of aesthetic experience. *Journal of Consciousness Studies, 6*(6–7), 15–51.

Rigau, J., Feixas, M., & Sbert, M. (2008). Informational aesthetics measures. *IEEE Computer Graphics and Applications, 28*(2), 24–34.

Rozin, P., & Royzman, E. B. (2001). Negativity bias, negativity dominance, and contagion. *Personality and Social Psychology Review, 5*(4), 296–320.

Schmidhuber, J. (1997). Low-complexity art. *Leonardo, 30*(2), 97–103.

Schmidhuber, J. (2007). *Simple algorithmic principles of discovery, subjective beauty, selective attention, curiosity & creativity* (Vol. 4754, pp. 32–33). Proceedings of the 18th international conference on Algorithmic Learning Theory (ALT), October 1–4, 2007, Sendai. Lecture Notes in Computer Science. arXiv:0709.0674.

Soofi, E. S., & Retzer, J. J. (2002). Information indices: Unification and applications. *Journal of Econometrics, 107*, 17–40.

Ware, C. (2008). *Visual thinking for design*. Burlington, MA: Morgan Kaufmann.

Williams, R. B. (2010). Why we love bad news. *Psychology Today*. Retrieved June 1, 2013, from http://www.psychologytoday.com/blog/wired-success/201012/why-we-love-bad-news. Accessed on March 10, 2014.

CHAPTER 2

Mental Models

2.1 MENTAL MODELS

Charles Perrow's book, *Normal Accidents: Living with High-Risk Technologies* (1984), collected a series of compelling cases that underline the role of human factors in many accidents. One of the cases was about an accident in Chesapeake Bay in 1978, involving a Coast Guard cutter training vessel. As they were traveling, the Coast Guard captain noticed another ship ahead in the darkness. He saw two lights on the other ship, so he thought that the ship was going in the same direction as his own ship. However, what he didn't realize was that there were actually three lights. That means it was moving toward them.

Since he missed one of the lights, his understanding of the situation was wrong. As both ships were traveling at full speed, they were closing in rapidly. Since the captain assumed they were going in the same direction, it seemed to him that their own ship was catching up with a very slow-moving fishing boat and that he was about to overtake the slow ship.

The other ship was, in fact, a large cargo ship. The two ships were heading toward each other. Both of them were approaching the Potomac River. The Coast Guard captain was still unaware that they were heading toward each other. Suddenly, he thought to make a left turn so that the small and slow fishing boat could turn to the port. Unfortunately, this decision resulted in the Coast Guard ship being put on a collision course with the oncoming freighter. Eleven coast guards on the ship were killed in the accident.

The captain's understanding of what was going on is his "mental model." He built his mental model on his initial observation of a ship with two lights instead of three. In his mental model, the two ships were traveling along the same direction. As new information became available, such as the distance between the two ships was closed up rapidly, he interpreted it using the framework provided by his mental model. On the Coast Guard ship, there was at least one other person who noticed the

The Fitness of Information: Quantitative Assessments of Critical Evidence, First Edition. Chaomei Chen.
© 2014 John Wiley & Sons, Inc. Published 2014 by John Wiley & Sons, Inc.

approaching cargo ship, according to Perrow's description. But there was no protocol for him to communicate what he saw to the captain. It was clear to him that the captain obviously noticed the ship as well; only he had no way to know that the captain got a wrong mental model of the situation.

The captain's mental model in this case was how he perceived, interpreted, and responded to the situation. He started with a wrong mental model and made a series of wrong decisions without questioning the validity of the model itself. As new information comes in, one can interpret and act with the current mental model, like what the captain did. In contrast, one may also challenge an established mental model, like how the OPERA's finding was interpreted.

Mental models are easy to form and hard to change. As the Coast Guard accident shows, one can come up with a detailed mental model based on a glimpse of the reality. Once a mental model is established, we tend to view the world through the lens of the new mental model. It becomes increasingly difficult to question the validity of a long-established mental model. Indeed, it may not even occur to us whether we should consider questioning the validity. On the other hand, research in several disciplines has shown that challenging an established model too soon or too often is not likely to be fruitful or productive either.

On a larger scale, mental models of the reality are formed not only individually, but also collectively. A shared mental model of the reality represents a consensus of a large number of people. For example, members of scientific communities may recognize and accept a common scientific theory such as Darwin's evolution and Einstein's relativity theories. Scholars may share a school of thought.

Thomas Kuhn's *Structure of Scientific Revolutions* is widely known for its introduction of the notion of a paradigm shift. A paradigm can be seen as the mental model of a scientific community. It functions similarly to the mental model of an individual. For scientists who work in a well-established paradigm, it may become increasingly difficult to adapt an alternative paradigm and see the world through a different lens. Kuhn used the term "Gestalt switch" to explain the challenge in changing from one perspective to another.

Is it possible to compare different mental models of the same underlying phenomenon? It may be similar to compare different versions of a story about the same event. Is it possible to tell which models, or paradigms, or stories, are superior? To whom is it superior? In what sense is it superior? These questions indicate that we should take into account two sources of information: one is our current mental model and the other is the external information, which could either be an isolated piece of information or a complex system of information, such as another mental model. If everything else is the same, a simpler model is likely to be superior for several reasons. For example, a simpler model is more efficient in terms of bridging the gap between the premises of the model and evidence that can be derived from the model. A simpler model is more plausible than a model that has to rely on a complex set of carefully crafted mechanisms. The Ptolemaic system of the solar system is a classic example. The Copernican model is a more accurate representation of the reality. The Copernican model is much simpler! Simplicity is one of the few criteria of a superior theory.

Even if the mental model is shared by a large number of intelligent people, it could still be a poor representation of the reality. Mental models and theories in general are about how the reality works or how a phenomenon takes place. They can be used to make predictions. Such predictions have direct influence on what decisions we make and which course of action we take. There are two broad categories for us to assess the quality of a mental model or a theory. We can examine the coherence and integrity of a theory internally. A theory is expected to explain the mechanisms of a phenomenon in a consistent manner. We can also examine the utility of a theory externally. Does it compete with an alternative theory? Can it give simpler explanations of the same phenomenon than its competitors?

2.1.1 Pitfalls

The intensity of our attention may drop dramatically as we move further away from the center of our attention. We tend to pay attention to various things that are immediately next to us but may overlook something important simply because it is further away. Similarly, the intensity of our interest in a topic decreases as the distance between the topic and topics that we are familiar with increases. This local tendency may cause problems. For instance, we tend to search locally, but the real solution to a problem may not be local. Searching beyond the local area, spatially or semantically, is often an inevitable step to find a better solution than the best solution available locally.

Another type of pitfall is that we tend to take paths of the least resistance rather than paths that are likely to lead to the best answers or the best decisions. The more often we go down the same path, the less resistance the path appears to be. Whenever a new path competes with a well-trodden one, our decision tends to be biased toward the familiar and proven path. We prefer to minimize uncertainty and want to avoid unforeseen risks. This preference may have serious consequences!

These pitfalls can be better explained in terms of mental models, or cognitive models. Mental models are simplified abstractions of how a phenomenon, the reality, or the world works. We use such models to describe, explain, and predict how things work and how situations may evolve. However, we are also biased by our own mental models.

As we have mentioned earlier, mental models are easy to form yet hard to change. Once we have established our mental model, we tend to see the world through the same mental model and reinforce it with new evidence. If we have to deal with evidence that apparently contradicts the model, our instinct is to find an interpretation for the contradiction rather than to question the model itself. Once we find extenuating reasons to convince ourselves why it makes sense that the evidence doesn't appear to fit the model, we move on with an uncompromised faith in the original model. Figure 2.1 shows a series of drawings that gradually change from a man's face to a sitting woman. If we start to look at these images one by one from the top left, we see images showing a man's face until the last few images. In contrast, if we start it from the lower right and move backward, we see more images showing a woman sitting there.

FIGURE 2.1 Mental models are easy to form but hard to change. Source: Unknown.

FIGURE 2.2 Connecting the dot with no more than four jointed straight lines.

Our perceptual ability enables us to recognize patterns easily from what we see. Since it comes so easy and effortless, we take the validity of these patterns for granted. Sometimes such patterns prematurely narrow down the solution space for subsequent search. Sometimes one may unconsciously rule out the ultimate solutions. A simple connecting-the-dot game illustrates this point.

In this game, nine dots are arranged in three rows and three columns (see Figure 2.2). You are asked to find a way to connect all these dots by four straight lines. The end point of each line must be the starting point of the next line.

If there is still no sign of a solution after a few trials, ask yourself whether you are making any implicit assumptions and whether you are imposing unnecessary constraints to your solutions. The problem is usually caused by implicit assumptions we make based on a Gestalt pattern that we may not even realize its existence. These implicit assumptions set the scope of our subsequent search. In this case, we won't be able to solve the problem unless the implicit assumptions are removed. This type of blind spot is more common in our thinking than we realize. Sometimes such blind spots are the direct source of accidents.

2.1.2 Communicating with Aliens

An effective communication test is to imagine how we could communicate an idea to aliens. In this scenario, we need to think carefully what we know about aliens and what they are supposed to know. This is an important element in communication—know your audience. Is it reasonable to assume that aliens know about something equivalent to Einstein's relativity theory? Can we count on it that aliens have a good knowledge of the Milky Way and our solar system? Is it necessary for us to explain to them who we are? The National Aeronautics and Space Administration's (NASA) scientists have demonstrated their answers to these questions in spacecraft *Pioneer* and *Voyager*.

Don Norman, a pioneer of human–computer interaction, proposed that one should check the reality through seven critical stages of an action cycle. Norman's action cycle starts with the intended goal to achieve, proceeds to the execution of a sequence of acts to reach the goal, and an evaluation of the effect of the action with respect to the original goal. Norman particularly identifies the execution and evaluation as the two most critical stages. Norman's suggestion was made primarily for human–computer interaction, especially in situations where an end user of a device or a system needs to figure out how the system is supposed to react to the user's action without knowing the exact design of the system. Based on the appearance and layout of various controls of the system and the results of initial trials of a few controls, the user may develop an idea how the system works. This type of understanding becomes the user's mental model. The focus of a mental model can be a situation as well as a device. September 11 and the mass destructive weapon are examples of incorrect mental models of situations at much larger scales (Betts, 2007).

The feedback from the system is critical for the user or the analyst to assess their mental model. For some systems, the analyst has to probe the systems to find out how the system would respond to particular input. The time elapsed for the user to receive feedback from the system is also critical, especially for interactive systems. If it takes too long for the user to receive feedback, it is less likely that the user would be able to make use of the feedback and refine the mental model. For many real-world systems, however, it is impossible to get any feedback, for example, on the long-term effects of genetically modified food on human beings and human activities on climate change.

A stabilized mental model is also known as a mindset. The stability comes with both pros and cons. The advantage is that a stable mental model provides us a familiar framework to solve routine problems or to apply the same methodology repeatedly. As we become more familiar with the framework, we become experts and specialize in this particular area of knowledge, and our performance becomes more efficient. However, a mindset is resistant to change. We often take our own mental model for granted. It becomes harder and harder for us to take a fresh perspective and question the validity of our own mental model than revise and modify our mental models. The most serious drawback of a mindset is that we become increasingly biased and less open minded. We tend to see everything through the same perspective, and we hardly question whether our mental model is appropriate in the first place.

2.1.3 Boundary Objects

The concept of *boundary objects* is useful for understanding and externalizing communications involving different perspectives (Bowker & Star, 2000; Star, 1989; Wenger, 1998). Boundary objects are artifacts that are common between different groups of people and flexible enough that each group may find more room to develop.

Boundary objects are externalized ideas. They form a shared context for communication. People can point to them in their communication as explicit reference points. The real value of a boundary object is its potential to stimulate communications and exchanges of ideas that neither parties have thought of beforehand.

We see what we expect to see. The same person may see different things in the same image at different times. Different people could see different things in the same image. Images taken by the Hubble Space Telescope mean different things to scientists and the public. It is a common practice that scientists make aesthetic enhancements or alterations to the original images before they are released to the public so that the pictures are easy to interpret. However, to astronomers, the "pretty pictures" that are intended for the public are qualitatively different from the purely "scientific images." To the public, these "pretty pictures" are taken as scientific rather than aesthetic. The *Pillars of Creation* was a famous example of such public-friendly pictures. The public not only treated it as a scientific image, but also attached additional interpretations that were not found in the original scientific image (Greenberg, 2004).

The Eagle nebula is a huge cloud of interstellar gas in the southeast corner of the constellation Serpens. Jeff Hester and Paul Scowen at Arizona State University took images of the Eagle nebula using the Hubble Space Telescope. They were excited when they saw the image of three vertical columns of gas. Hester recalled, "we were just blown away when we saw them." Then their attention was directed to "a lot of really fascinating science in them" such as the "star being uncovered" and "material boiling" off of a cloud.

Greenberg (2004) described how the public reacted to the image. The public first saw the image on CNN evening news. CNN received calls from hundreds of viewers who claimed that they have seen apparition of Jesus Christ in the Eagle nebula image. On CNN's live call-in program next day, according to viewers, they were able to see more: a cow, a cat, a dog, Jesus Christ, the Statue of Liberty, and Gene Shalit (a prominent film critic).

The reactions to the Eagle nebula image illustrate the concept of a *boundary object*, which is subject to reinterpretation by different communities. A boundary object is both vague enough to be adopted for local needs and yet robust enough to maintain a common identity across sites. Different communities, including astronomers, religious groups, and the public, put various meanings to the image. More interestingly, nonscientific groups were able to make use of the scientific image and its unchallenged absolute authority for their own purposes so long as newly added meanings do not conflict with the original scientific meaning. Greenberg's analysis underlines that the more the scientific process is black boxed, the easier it becomes to augment scientific knowledge with other extrascientific meanings.

2.1.4   Wrong Models

The Coast Guard accident mentioned at the beginning of the chapter is obviously an example of a wrong mental model and consequences of having a wrong model. As we will illustrate with a few more examples in this section, having a wrong mental model has something to do with how we process incoming information, what assumptions we make when we process such information, and how we would know when we should reconsider and validate an assumption that we have been taking it for granted for a long time.

The Wrong Patient

Mark R. Chassin and Elise C. Becher (2002) revealed a case in which a patient was mistaken for another patient who needed an invasive electrophysiology procedure. Their study highlighted that although there were at least 17 distinct errors, none of them alone could have caused the problem. This example underlines how often we make assumptions when critical information is missing, how strongly our views may be biased by our current mental model, and how hard it is to realize that a mental model could be wrong. We summarize what happened based on Chassin and Becher's detailed descriptions and discuss missed opportunities that one could recognize the mistake earlier on.

The first patient is a 67-year-old woman named Joan Morris (a pseudonym). She is a native English speaker and high school graduate. After a head injury, she was admitted to a teaching hospital for cerebral angiography, a procedure that uses a special dye and x-rays to see how blood flows through the brain. After the procedure, she was transferred to the oncology floor rather than the telemetry unit. She was supposed to be discharged the following day, but it was not clear why she was not sent to the telemetry unit. In hindsight, it was the first error that could have been avoided.

The second patient is a 77-year-old woman with a similar name Jane Morrison (a pseudonym). She was transferred to the hospital for a cardiac electrophysiology procedure. She was admitted to the telemetry unit, the same unit as the first patient. The electrophysiology procedure was scheduled for the morning next day.

The next morning, the electrophysiology nurse (RN1) telephoned the telemetry unit and asked for "patient Morrison." RN1 was falsely informed that Ms. Morrison had been moved to the oncology floor. In fact, Ms. Morrison was still there in the telemetry unit. This was the second error and probably the most critical one because the wrong information made the nurse to chase the wrong patient.

The oncology floor told RN1, wrongly, that her patient was there and would be sent to the electrophysiology laboratory. So far, three different sources gave the nurse RN1 the wrong information. The reason appears that two names are so similar. This also means that this type of error could easily happen without anyone realizing it.

The first patient was supposed to be discharged in the morning and not to have any procedure at all. Her nurse RN2 was asked to send her over for the electrophysiology procedure. Routinely, there should be a written order for the procedure, but there was no written order. It was the end of RN2's night shift. She assumed the procedure was scheduled despite the absence of a written order. The patient told RN2 that she was unaware of plans for an electrophysiology procedure and she didn't want to do it. RN2 told the patient that she could refuse it after she arrived in the laboratory. Here is another missed opportunity. Instead of making an assumption to explain away the absence of an order, if RN2 could investigate further about the order and connect to the fact that the patient herself was not aware of the procedure, there might be a chance to stop the chain of errors from going further.

Once in the lab, the patient told them again that she was reluctant to go through the procedure. RN1 paged the attending physician, their supervisor. The attending physician met the right patient the night before, but didn't realize he was speaking to a different person over the phone. He stated to RN1 that the patient had agreed to proceed. In hindsight, the patient expressed her reluctance, but probably did not stress that she was not aware of the procedure. The nurse RN2 probably did not mention that the writing order was missing. A potentially more effective strategy for the patient to convince the attending physician is probably to draw the attending physician's attention to the incomplete paperwork. On the other hand, if it didn't occur to the patient that doctors could pick the wrong patient, then it is less likely for the patient to consider how to deal with the issue directly.

The (right) patient's nurse RN1 noticed there was no consent form. She paged the fellow who scheduled the procedure. The fellow came and was surprised. He then discussed with the (wrong) patient and had her sign the consent form. Here a major opportunity was missed. The role of a consent form was degenerated to a formality. It was not clear whether the hospital had a more stringent protocol to deal with situations like this so that the patient had a chance to truly understand what she was signing for. The authors mentioned that the patient was a high school graduate, and they questioned what made her give her consent on a procedure that she was not aware of and didn't want to proceed with.

The next opportunity came when the patient's resident physician was surprised to find his patient was taken to a procedure that he did not order, but after discussing it with RN1, he assumed the attending physician had ordered the procedure without informing him. Let's look at what RN1 knew up to this point. She knew that the patient was reluctant to go through the procedure, the patient didn't have a consent form, and the patient's resident physician was unaware of the procedure. All the information apparently didn't draw the attention of the nurse RN1 to the possibility that something was probably wrong.

Another unnoticed signal came from a nurse from the telemetry floor who checked with the electrophysiology laboratory on why the second patient, the right patient, was not called for her procedure. In the meantime, the electrophysiology charge nurse noticed that the patient's name (the first and the wrong patient) was not on the schedule list for the procedure. She raised the question to the attending physician and she was told "This is our patient." She then assumed that the patient was added to the list after the schedule had been distributed. When human beings encounter information that appears to be inconsistent with their mental models, a common way to resolve the inconsistency is to "explain it away" by finding a reason that could explain the inconsistency. Such "convenient" explanations are often ad hoc to serve the sole purpose of filling an apparent gap. More dangerously, their validity is rarely questioned further. Instead, they are often taken for granted and the focus of the reasoning moves quickly to the next issue. The quick response from the attending physician, "This is our patient," indicates that he probably didn't even think it may be necessary to justify why this was indeed the right patient. The nurse was able to convince herself by explaining away the missing name from the schedule.

The final opportunity came from a radiology attending physician who couldn't find the first patient and asked the electrophysiology laboratory, specifically, why Ms. Morris was having this procedure. The electrophysiology attending physician stated to the nurse that the call was about Ms. Morris, but the patient on the table was Ms. Morrison. The nurse corrected him, stating that in fact the patient on the table was Morris, not Morrison. The attending physician finally recognized the error. A few lessons can be learned from here. The error was revealed because specific names were mentioned by the x-ray physician and by the attending physician so that the nurse had a chance to inform the attending physician specifically where the mismatch was. In several earlier missed opportunities, the patient's name was not explicitly communicated. Instead, "our patient" and "patient Morrison" were used. The former was vague and the second was easily mistaken as Morris. Many assumptions were made to eliminate the apparent gaps between what was expected and what was experienced. Unfortunately, the validity of these assumptions was never questioned nor verified.

After a series of errors, the mistake was finally noticed. The patient was one hour into the procedure. She returned to her room in stable condition. This example vividly illustrates the consequences of having a wrong mental model. Unlike the Coast Guard captain, many people were involved in the wrong patient case. Many assumptions were made by individuals without explicitly communicating to others. When information was not consistent with their mental models, in many occasions, people explained away the discrepancies with little or no justification.

While sometimes a single piece of information may change our beliefs altogether, there are also numerous occasions where we simply ignore any information that is particularly intended to change our beliefs and our behaviors. Both physicians and patients are often found not to adhere to what they are supposed to do in terms of guidelines or instructions to take medications.

Clinical practice guidelines are systematically developed to inform practitioners and physicians about appropriate health care for specific clinical circumstances. However, research has found that guidelines do not have a considerable impact on physician behavior. Cabana and his colleagues reviewed barriers to physician adherence to clinical practice guidelines. They reviewed 76 studies selected from a larger sample and identified 293 barriers to adherence to clinical guidelines.

Their survey revealed that the adherence problem was complex and barriers vary considerable from one setting to another. Common barriers include lack of awareness, lack of familiarity, lack of agreement, lack of outcome expectancy, lack of self-efficacy, and inertia of previous practice. In a broader context, diffusion of innovations has been used in the study of the adoption of clinical guidelines as well as the spread of innovative technologies. Diffusion of innovations in general is a multistep process involving the awareness of the knowledge or recommended guidelines, persuaded by forming an opinion about the recommendation, and decision making to determine whether or not to adopt a recommendation.

Taken together, it is our opinion that every error in the wrong patient case identifies an opportunity missed. Although we agree that each single error alone could not cause the entire problem, preventing any of the errors would significantly reduce the risk for the system failure as a whole. A well-known model of this type of system failure is called the Swiss cheese model. A complex system consists of multiple layers of cheese with holes on each layer. As long as these holes do not align up a hole that goes through all the layers simultaneously, then the system can tolerate these local imperfections. The question is then: what is the likelihood of these holes lining up?

Connecting the Right Dots

Why do we often jump to conclusions? One possible reason is that we tend to underestimate the complexity of the nature or the world in general. If two events took place one after the other, we tend to assume that the first event somehow caused the second one. When we need to find an explanation of what is going on, we often settle with the first good enough reason that we can find. In reality, in both types of situations, we may just pick the wrong end of the stick, so to speak.

As an urban legend says, a family complained to General Motors' Pontiac Division about a strangely behaved engine of their new Pontiac. The family had a tradition of having ice cream for dessert after dinner every night. The whole family would vote on which flavor of ice cream and the father would drive his new Pontiac to the store to get it. Strangely enough, trips for vanilla ice cream always ran into the same problem—the car won't start on the way back, but if he gets any other kind of ice cream, the car would start just fine.

Pontiac sent an engineer to check it out. The engineer rode the car with a few ice cream trips. The car stalled on vanilla ice cream trips and started on other flavors, just the way it was complained. Was the car allergic to vanilla ice cream?

The engineer noted a variety of details of the trips such as the type of gas used and time to drive back. Then, he had a clue: It took less time to buy vanilla than any other flavor. Because vanilla ice cream is popular, the store puts it at the front of the store and it is easy to get. It takes much longer to get other flavored ice cream because it is located in the back of the store. Now the question for the engineer was why the car wouldn't start when it took less time. The engineer was able to locate the problem. It was with the vapor lock. The engine needs enough time to cool down.

Richard Betts is a former member of the Military Advisory Panel of the Director of Central Intelligence and of the National Commission on Terrorism. He compared two cases of intelligence failure—September 11 and Iraq's missing weapon of mass destruction (WMD): in one case, the intelligence community failed to provide enough warning; in the other, it failed by providing too much (Betts, 2007).

It was commonly believed after the September 11th terrorist attacks that U.S. intelligence had failed badly. However, Betts pointed out the issue is not that simple. The intelligence system did detect that a major attack was imminent weeks before the

September 11 attacks. The system warned clearly that an attack was coming, but could not say where, how, or exactly when. The vital component lacking from the warning was the actionability—it was too vague to act upon.

According to Betts, two months before September 11th, an official briefing warned that Bin Laden "will launch a significant terrorist attack against U.S. and/or Israeli interest in the coming weeks. The attack will be spectacular and designed to inflict mass casualties." George Tenet, the Director of Central Intelligence, later said in his memoirs that "we will generally not have specific time and place warning of terrorist attacks." In addition, many intercepted messages or cryptic warnings were not followed by any terrorist attack. Before September 11, more than 30 messages had been intercepted, and there was no terrorist attack. Furthermore, it is not unusual to choose not to act on warnings if various uncertainties are involved. An extreme hurricane in New Orleans had been identified by the Federal Emergency Management Agency (FEMA) long before the Hurricane Katrina arrived in 2005. Fixing New Orleans's vulnerability would have cost an estimated US $14 billion. The perceived cost-effectiveness before the disaster was not in favor of making such investments while its benefit was hard to estimate. The question sometimes is not how to act upon a credible assessment of a threat or a potential disaster; rather, it is prudent to ask whether one should act given the cost-effectiveness in the situation. Gambling sometimes pays off. Other times, it will be seen as a failure, especially in hindsight. In Chapter 4, we discuss the gambling nature of almost all decision making in terms of the optimal foraging framework.

Another factor that contributed to the failure was due to the loss of focus caused by the tendency of maximizing collection. The trade-off between collecting more dots and connecting the dots is the issue. The fear of missing any potentially useful dots and the fear of taking direct responsibilities were driving the maximum collection of dots. However, collecting more dots makes connecting the dots even harder. Indeed, after reading the 9/11 Commission's report, Richard Posner concluded that it is almost impossible to take effective action to prevent something that has never occurred previously. In many ways, this is a question also faced by scientists, who are searching for ways to find meaningful dots and connect them. As we will see in later chapters, there are pitfalls as well as effective strategies for dealing with such situations.

The 9/11 Commission recommends that the dots should be connected more creatively and it is "crucial to find a way of routinizing, even bureaucratizing, the exercise of imagination." Mechanisms for thinking outside the box should be promoted and rewarded, just as Heuer (1999) recommended.

If American intelligence failed to connect the dots before September 11, it made the opposite mistake on Iraq by connecting the dots too well. Betts showed that ironically policymakers paid little attention to intelligence on cultural and political issues where the analysis was right, but they acted on technical intelligence about weapons of mass destruction, which was wrong.

The consensus after the first Gulf War was that Iraq had attempted to develop WMD. The lack of evidence that they had indeed destroyed the necessary facilities was interpreted as a sign that they were hiding from the inspections of the West.

The mindset of the intelligence and policymakers was that the Iraqis concealed WMD as they did with their chemical and biological weapons prior to the first Gulf War. The lack of direct evidence of the existence only reinforced the model rather than raised an alarm about the assumption. After all, the lack of evidence was consistent with the hypothesis that Saddam was hiding it from the world. In addition, negative evidence that pointed to different scenarios did not get enough attention. It is known from psychology that people tend to maintain their existing mental models rather than challenge them when facing new evidence. It takes much stronger evidence, that is, wake-up calls, to make people alter their mental model. In this case, analysts did not ask whether Iraq had WMD; instead, they wanted evidence that Iraq did not have WMD.

Given this mindset, conclusions were drawn from Iraq's behavior in the past rather than from any direct evidence of the existence from the present. For instance, documents obtained by the United Nations inspectors showed that an Iraqi government committee once gave instructions to conceal WMD activities from inspectors. At the end of the first Gulf War, Iraq admitted having chemical and biological weapons and claimed that they had destroyed them later, but never produced any evidence of such destructions. Misinterpretations of all available evidence and the lack of evidence only became obvious in hindsight. Dots were connected without evidence beyond the doubt. True connections were slipped through the analysis.

In hindsight, Betts suggested what the intelligence should have done, given what was known and what was possible to know at the time. First, Iraq was probably hiding WMD weapons. Second, the sources that led to the conclusion were deductions from the past history. And third, there was very little direct evidence to back up the deduction. One of the lessons learned from the two opposite cases is a caution that one should not draw too many lessons from a single failure.

Pearl Harbor: Warning and Decisions by Roberta Wohlstetter (1962) was regarded as the first intelligence analysis that differed significantly from prior studies of surprise attacks. Wohlstetter's focus was on Pearl Harbor, but his insight is far-reaching. He studied Pearl Harbor in a much broader context than previous studies of similar surprise attacks would reach. The key point he made was that analysts and decision makers in crisis situation like Pearl Harbor can be seriously biased by our own perception or, rather, misperception. The Pearl Harbor surprise was not due to a lack of relevant intelligence data. Instead, it was due to misperceptions of the available information.

Abraham Ben-Zvi (1976) extended this line of analysis of surprise attacks and paid particular attention to both strategic and tactical dimensions. He analyzed five cases of intelligence failure to foresee a surprise attack and found that whenever strategic assumptions and tactical indicators of attack converged, an immediate threat was perceived and appropriate precautions were taken. When discrepancies emerged, people chose to rely on strategic assumptions and discard tactical indicators as noises.

The weaknesses of human perception and cognition have been identified in numerous cases across a diverse range of situations. Nevertheless, we still need to promote the awareness of these weaknesses and remind ourselves that conflicting information may be an early sign for reassessing what we think we know about a situation. The "noises" may contain vital clues. As Heuer (1999) pointed out, it is essential to be able to recognize when our mental model needs to change.

Rejecting Good Ideas

The Nobel Prize is widely regarded as the ultimate recognition of one's outstanding achievements in physics, chemistry, medicine, literature, and peace. In his will, Alfred Nobel described the prizes should be given to persons who have made the most important discoveries and the most outstanding work or who have done the best work for promoting peace regardless of their nationality:

> one part to the person who shall have made the most important discovery or invention within the field of physics; one part to the person who shall have made the most important chemical discovery or improvement; one part to the person who shall have made the most important discovery within the domain of physiology or medicine; one part to the person who shall have produced in the field of literature the most outstanding work in an ideal direction; and one part to the person who shall have done the most or the best work for fraternity between nations, for the abolition or reduction of standing armies and for the holding and promotion of peace congresses. (http://www.nobelprize.org/alfred_nobel/will/will-full.html)

Although there is no question that the prizes should be given to the most important discoveries and the most outstanding work, there are always discrepancies, to say the least, on which discovery is the most important and which work is the most outstanding. While the importance of some of the Nobel Prize-winning work was recognized all along, the significance of other Nobel Prize-winning work was overlooked or misperceived. How often do experts recognize the potential of important work? How often do they miss it? Why and how?

Peer review is a long-established tradition in science. Scientists publish papers. Scientists seek funding to support their work and submit grant proposals to funding authorities. Peer review plays the most critical role in both scientific publication and funding allocation. There is a consensus, at least in the scientific community, that peer-reviewed publications or grant proposals have a more prestigious status and a better quality than non-peer-reviewed ones. In the peer review system, scientists submit their work for publication. The decision on whether their work should be published heavily relies on the outcome of reviews produced by peer scientists who usually work in the same field of study. Reviewers usually make recommendations whether a manuscript is publishable and whether additional changes are necessary before they get published. Reviewers and authors may or may not be anonymous to each other. The anonymity can be either one way or two way, known as a single-blind or a double-blind process, respectively. Double-blind peer reviews are regarded as more rigorous than other forms of peer reviews. Peer review of grant proposals works similarly.

As one can imagine, peer reviews have made blunders in both ways: blocking the good ones and letting the bad ones through. Tim Berners-Lee's paper on the idea that later led to the World Wide Web was rejected by the ACM Hypertext conference on the grounds that it was too simple by the standard of the research community. Reviewers of grant proposals were criticized to be too conservative to judge transformative research and high-risk and high-payoff proposals.

Rejecting a future Nobel Prize-winning discovery would be a big enough blunder to get everyone concerned. If an achievement can be ultimately recognized by a Nobel Prize, how could experts misjudge its potential in its early days?

One of the most commonly given reasons for rejection in general is that the work in question is premature. Spanish scholar Juan Miguel Campanario noted a lack of interest from sociologists, philosophers, and historians of science on how and why some important scientific discoveries were rejected, resisted, or simply ignored by peer scientists. Keeping an open mind is one of the most fundamental values upheld by a scientist. However, reviewers must also consider risks and uncertainties. In order to better understand what transformative research is and how one can recognize it timely, it is important to understand not only how scientific breakthroughs get instant recognition but also how revolutionary discoveries can be rejected, resisted, and ignored. Campanario identified some common patterns of resistance to scientific discovery:

- Papers are rejected.
- Discoveries are ignored by peer scientists.
- Published papers are not cited.
- Commentaries oppose new discoveries.

Campanario analyzed commentaries written by authors of highly cited papers and found that some of the most cited papers experienced initial rejections. It is an interesting phenomenon that some of the rejected papers became highly cited later on. Campanario then studied 36 cases of rejected Nobel class articles and 27 cases of resisted Nobel class discoveries and concluded that there is a real danger and the consequence can be disastrous. He searched through autobiographies of Nobel Prize winners and conducted a survey among Nobel laureates awarded between 1980 and 2000 and received 37 personal accounts of experiences of resistance. Resistance was found in two categories: skeptics toward a discovery that ultimately received a Nobel Prize and rejections of papers reporting a discovery or a contribution that was later awarded with a Nobel Prize. Campanario located passages from autobiographies and personal accounts to illustrate the nature of rejection or resistance. I added brief contextual information to the following examples for clarity.

The Nobel Prize in Physiology or Medicine 1958 was awarded jointly to George Wells Beadle and Edward Lawrie Tatum "for their discovery that genes act by regulating definite chemical events" and the other half to Joshua Lederberg "for his discoveries concerning genetic recombination and the organization of the genetic material of bacteria." In his 1974 recollections, Beadle described that "… few people were ready to accept what seemed to us to be a compelling conclusion" (Beadle, 1974).

The Nobel Prize in Physics 1964 was awarded jointly to Charles Hard Townes, Nicolay Gennadiyevich Basov, and Aleksandr Mikhailovich Prokhorov "for fundamental work in the field of quantum electronics, which has led to the construction of oscillators and amplifiers based on the maser-laser principle." Townes recalled the pressure from peers who themselves were Nobel laureates: "One day... Raby and Kusch, the former and current chairmen of the department, both of them Nobel Laureates for their work with atomic and molecular beans and with a lot of weight behind their opinions, came into my office and sat down. They were worried. Their research depended on support from the same source as did mine. 'Look', they said, 'you should stop the work you are doing. You're wasting money. Just stop'." (Lamb, Schleich, Scully, & Townes, 1999, p. S266)

The second category of resistance identified by Campanario is rejections of Nobel class papers. He acknowledged that some of the rejections were indeed justified and in some cases the Nobel Prize-winning versions differ from the initial versions.

Nobel laureate Allan Cormack's publications in the well-known *Journal of Applied Physics* in 1963 and 1964 were examples of so-called sleeping beauties, that is, publications with delayed recognition. These articles introduced his theoretical underpinnings of computed tomography (CT) scanning. However, the two articles generated little interest—they attracted seven citations altogether for the first 10 years—until Hounsfield used Cormack's theoretical calculations and built the first CT scanner in 1971. Cormack and Hounsfield shared the 1979 Nobel Prize in Physiology or Medicine.

Nobel laureate Stanley Prusiner wrote, "while it is quite reasonable for scientists to be skeptical of new ideas that do not fit within the accepted realm of scientific knowledge, the best science often emerges from situations where results carefully obtained do not fit within the accepted paradigms." Prusiner's comment echoes our earlier discussions on the potential biases of one's mindset. Both researchers and reviewers are subject to such biases. While it is reassuring to know that some early rejected works do become recognized later on, it is hard to find out how many, if any, potentially important discoveries discontinued because their values were not recognized soon enough.

2.1.5   Competing Hypotheses

We are often torn by competing hypotheses. Each hypothesis on its own can be very convincing, while they apparently conflict with each other. One of the reasons we find hard to deal with such situations is because most of us cannot handle the cognitive load needed for actively processing multiple conflicting hypotheses simultaneously. We can focus on one hypothesis, one option, or one perspective at a time. In literature, the commonly accepted magic number is 7, plus or minus 2. In other words, if a problem involves about five to nine aspects, we can handle them fine. If we need to do more than that, we need to, so to speak, externalize the information, just as we need a calculator or a piece of paper to do multiplications beyond single-digit numbers.

It is much easier to convince people using vivid, concrete, and personal information than using abstract, logical information. Even physicians, who are well qualified to understand the significance of statistical data, are convinced more easily by vivid personal experiences than by rigorous statistical data. Radiologists who examine lung x-rays everyday are found to have the lowest rate of smoking. Similarly, physicians who diagnosed and treated lung cancer patients are unlikely to smoke.

Analysis of competing hypotheses (ACH) is a procedure to assist the judgment on important issues in a complex situation. It is particularly designed to support decision making involving controversial issues by keeping track what issues analysts have considered and how they arrived at their judgment. In other words, ACH provides the provenance trail of a decision-making process.

The ACH procedure has eight steps (Heuer, 1999):

1. Identify the possible hypotheses to be considered. Use a group of analysts with different perspectives to brainstorm the possibilities.
2. Make a list of significant evidence and arguments for and against each hypothesis.
3. Prepare a matrix with hypotheses across the top and evidence on the side. Analyze the "diagnosticity" of the evidence and arguments—that is, identify which items are most helpful in judging the relative likelihood of the hypotheses.
4. Refine the matrix. Reconsider the hypotheses and delete evidence and arguments that have no diagnostic value.
5. Draw tentative conclusions about the relative likelihood of each hypothesis. Proceed by trying to disprove the hypotheses rather than prove them.
6. Analyze how sensitive your conclusion is to a few critical items of evidence. Consider the consequences for your analysis if that evidence were wrong, misleading, or subject to a different interpretation.
7. Report conclusions. Discuss the relative likelihood of all the hypotheses, not just the most likely one.
8. Identify milestones for future observation that may indicate events are taking a different course than expected.

The concept of diagnostic evidence is important. The presence of diagnostic evidence removes all the uncertainty in choosing one hypothesis over an alternative. Evidence that is consistent with all the hypotheses has no diagnostic value. Many illnesses may have the fever symptom, thus fever is not diagnostic evidence on its own. In the mass extinction example, the analogy of the KT mass extinction does not have sufficient diagnostic evidence to convince the scientific community. We use evidence in such situations to help us estimate the credibility of a hypothesis.

Risks and Payoffs

A turning point in U.S. science policy was during 1967 and 1968. Up until then, science policy had been dominated by the cold war. By 1963, the national investment in R&D was approaching 3% of GDP, 2.5 times the peak reached just before the end of World War II. More than 70% of this effort was supported by the federal government. Ninety-three percent of it came from only three federal agencies: the Department of Defense

(DoD), the Atomic Energy Commission, and NASA. Their mission was to ensure the commercial spinoff of the knowledge and technologies they had developed. Much of the rest of federal R&D was in the biomedical field. Federal funding for basic research in universities reached a peak in 1967. It declined after that in real terms until about 1976.

The current funding environment is very competitive because of the tightened budget and increasing demands from researchers. In addition, the view that science needs to serve the society means that funding authorities as well as individual scientists need to justify how scientific inquiries meet the needs of society in terms of economic, welfare, national security, and competitiveness. There are two types of approaches to assess the quality and impact of scientific activities and identify priority areas for strategic planning. One is qualitative in nature, primarily based on opinions and judgments of experts, including experts in scientific fields, specialists in relevant areas of applications, and end users. The other is quantitative in nature, mainly including the development and use of quantitative metrics and indicators to provide evidence for making assessments and decisions. Metrics and indicators that can rank individual scientists, institutions, countries, as well as articles and journals have become increasingly popular. *Nature* recently published a group of featured articles and opinion papers on the issue of assessing assessment (Editorial, 2010).

Funding agencies have been criticized for their peer review systems for being too conservative and reluctant to support high-risk and unorthodox research. Chubin and Hackett (1990) found 60.8% of researchers supported this notion and 17.7% disagreed. Peer review has been an authoritative mechanism used to select and endorse research proposals and publications. Chubin and Hackett describe peer review as an intensively private process: it originates within a scientist's mind, continues on paper as a bureaucratic procedure, and ends behind the closed doors of a funding agency.

Laudel (2006) studied how researchers in Germany and Australia adapt their research in response to the lack of recurrent and external funding. She also confirmed the perception that mainstream research is a key. One scientist said that he would not send a grant proposal unless he has at least two publications in the same area. She also noted the switch from basic research to applied research. Interdisciplinary research is also among the types of research that suffer from the tightened funding climate.

The profound problem in this context is closely related to the one that has been troubling the intelligence community. There is simply too much information and yet too little salient and meaningful signs. In addition, data cannot speak for itself. We need to have theories and mental models to interpret data and make sense patterns emerged from data. There are many theories of how science evolves, especially from the philosophy of science, but these theories are so different that they explain the same historical event in science in totally different perspectives.

A considerable amount of inventions are built upon previously known technological features. Hsieh (2011) suggested an interesting perspective to see new inventions as a compromise between two contradictory factors: the usefulness of an invention and the relatedness of prior inventions that the new invention is to synthesize. He referred to such prior inventions as features. If these features are closely related, then the inventor will have to spend a lot of effort to distinguish them. On the other hand, if these features are barely related, the inventor will have to find out how they might be related.

The optimal solution would be somewhere between the two extreme situations. The relations between features should neither be too strong nor too weak. The most cost-effective strategy for the inventor is to minimize the high costs of connecting unrelated features and simultaneously minimize the costs of synthesizing ones that are already tightly connected. Hsieh tested his hypothesis with U.S. patents granted between 1975 and 1999. The usefulness of an invention was measured by future citation, while the relatedness of inventions was measured by the network of citations. He found a statistically significant inverse U-shaped relationship between an invention's usefulness and the relevance among its component features. The usefulness of an invention was relatively low when the relevance was either too strong or too weak. In contrast, the usefulness was the highest when the relatedness was in between the two extremes.

Transformative research is often characterized as being high risk and potentially high payoff. Revolutionary and groundbreaking discoveries are hard to come by. What are the implications of the trade-off strategy on funding transformative research with public funding? It is known that it takes long time before the values of scientific discoveries and technological innovations become clear. Nobel Prize winners are usually awarded for work that they conducted a few decades ago. We also know that Nobel class ideas do get rejected. The question is: to what extent will we be able to foresee scientific breakthroughs?

Project Hindsight

How long does it take for the society to fully recognize the value of scientific breakthroughs or technological innovations? *Project Hindsight* was commissioned by the U.S. DoD in order to search for lessons learned from the development of some of the most revolutionary weapon systems. One of the preliminary conclusions drawn from *Project Hindsight* was that basic research commonly found in universities didn't seem to matter very much in these highly creative developments. It appears, in contrast, that projects with specific objectives were much more fruitful.

In 1966, a preliminary report of *Project Hindsight* was published. A team of scientists and engineers analyzed retrospectively how 20 important military weapons came along, including Polaris and Minuteman missiles, nuclear warheads, C-141 aircraft, Mark 46 torpedo, and the M 102 Howitzer. Researchers identified 686 "research or exploratory development events" that were essential for the development of the weapons. Only 9% were regarded as "scientific research" and 0.3% was base research. Nine percent of research was conducted in universities.

Project Hindsight indicated that science and technology funds deliberately invested and managed for defense purposes have been about one order of magnitude more efficient in producing useful events than the same amount of funds invested without specific concern for defense needs. *Project Hindsight* further concluded that:

1. The contributions of university research were minimal.
2. Scientists contributed most effectively when their effort was mission oriented.
3. The lag between initial discovery and final application was shortest when the scientist worked in areas targeted by his sponsor.

In terms of its implications on science policy, *Project Hindsight* emphasized mission-oriented research, contract research, and commission-initiated research. Although these conclusions were drawn from the study of military weapon development, some of these conclusions found their way to the evaluation of scientific fields such as biomedical research.

The extended use of preliminary findings had drawn considerable criticism. Comroe and Dripps (2002) criticized *Project Hindsight* as anecdotal and biased, especially because it was based on the judgments of a team of experts. In contrast to the panel-based approach taken by *Project Hindsight*, they started off with clinical advances since the early 1940s that have been directly responsible for diagnosing, preventing, or curing cardiovascular or pulmonary disease, stopping its progression, decreasing suffering, or prolonging useful life. They asked 40 physicians to list the advances they considered to be the most important for their patients. Physicians' responses were grouped two lists in association with two diseases: a cardiovascular disease and a pulmonary disease. Then, each of the lists was sent to 40–50 specialists in the corresponding field. Specialists were asked to identify corresponding key articles, which needed to meet the following criteria:

1. It had an important effect on the direction of subsequent research and development, which in turn proved to be important for clinical advance in one or more of the 10 clinical advances they were studying.
2. It reported new data, new ways of looking at old data, new concept or hypothesis, a new method, new techniques that either was essential for full development of one or more of the clinical advances or greatly accelerated it.

A total of 529 key articles were identified in relation to 10 advances in biomedicine:

1. Cardiac surgery
2. Vascular surgery
3. Hypertension
4. Coronary insufficiency
5. Cardiac resuscitation
6. Oral diuretics
7. Intensive care
8. Antibiotics
9. New diagnostic methods
10. Poliomyelitis

It was found that 41% of these advances judged to be essential for later clinical advance were not clinically oriented at the time they were made. The scientists responsible for these key articles sought knowledge for the sake of knowledge. Of the key articles, 61.7% described basic research; 21.2% reported other types of research; 15.3% were concerned with the development of new apparatus, techniques, operations, or procedures; and 1.8% were review articles or synthesis of the data of others.

Comroe and Dripps discussed research on research, similar to the notion of science of science. They pointed out that it requires plenty of time and support to conduct research, both retrospective and prospective, on the nature of scientific discovery and to understand the courses of long and short lags between discovery and application. Their suggestion echoes the results of an earlier project commissioned by the NSF in response to *Project Hindsight*. NSF argued that the time frame studied by the *Hindsight* project was too short to identify the basic research events that had contributed to technological advances. The NSF commissioned a project known as TRACES to find how long it would take for basic research to evolve to the point that potential applications become clear. However, Mowery and Rosenberg (1982) argued that the concept of research events is much too simplistic and neither *Hindsight* nor TRACES used a methodology that is capable enough of showing what they purport to show.

TRACES

TRACES stands for *Technology in Retrospect and Critical Events in Science*. It was commissioned by the NSF. *Project Hindsight* reviewed the history of 20 years prior to the inventions, but TRACES examined the history of five inventions and their origins dated back as early as the 1850s. The five inventions are the contraception pill, matrix isolation, the videotape recorder, ferrites, and the electron microscope. TRACES identified 340 critical research events associated with these inventions and classified these events into three major categories: nonmission research, mission-oriented research, and development and application. Seventy percent of the critical events were nonmission research, that is, basic research. 20% were mission oriented, and 10% was development and application. Universities were responsible for 70% of nonmission and 30% of mission-oriented research. For most inventions, 75% of the critical events occurred before the conception of the ultimate inventions.

Funding agencies have many reasons to evaluate their research portfolios. They need to justify their funding decisions to the congress, the general public, and scientists. Strategically, there are three sets of requirements: the long term (10–20 years), the short term (3–5 years), and the near term (1–2 years). As funding agencies receive more proposals at an increasing rate, they also increasingly recognize the fundamental role of transformative research and other high-risk, high-payoff research in science and technology. In Chapter 5, we will discuss this issue from the perspective of optimal foraging in terms of maximizing the ratio of expected returns to risks.

The point of invention of the videotape recorder was the mid-1950s. It took almost 100 years to complete 75% of all relevant events, but the remaining 25% of the events converged rapidly within the final 10 years. In particular, the conception to innovation period took place in the final five years.

The invention of the videotape recorder involves six areas: control theory, magnetic and recording materials, magnetic theory, magnetic recording, electronics, and frequency modulation. The earliest nonmission research appeared in the area of magnetic theory: Weber's early ferromagnetic theory in 1852. The earliest mission-oriented research appeared in 1898 when Poulsew used steel wire for the first time for recording. According to TRACES, the technique was "readily available but had many basic limitations, including twisting and single track restrictions." Following Poulsew's work in the areas of Magnetic and Recording Materials, Mix & Genest was able to develop steel tape with several tracks around the 1900s but limited by lack of flexibility and increased weight. This line of invention continued as homogeneous plastic tape on the magneto-phon tape recorder was first introduced in 1935 by AEG. A two-layer tape was developed by the 1940s. The development of reliable wideband tapes was intensive in the early 1950s. The first commercial videotape recorder appeared in the late 1950s.

The invention of electron microscope went through similar stages. The first 75% of research was reached before the point of invention, and the remaining research consisted of a translational period from conception to innovation.

The invention of electron microscope relied on five major areas, namely, cathode ray tube development, electron optics, electron sources, wave nature of electrons, and wave nature of light. Each area may trace several decades back to the initial nonmission discoveries. For instance, Maxwell's electromagnetic wave theory of light in 1864, Roentgen's discovery of emission of x-ray radiation in 1893, and Schrodinger's foundation of wave mechanics in 1926 all belong to nonmission research that ultimately led to the invention of electronic microscope. TRACES reveals that between 1860 and 1900 there were almost no connections across these areas of nonmission research. While the invention of electronic microscope was dominated by many earlier nonmission activities, the invention of the videotape recorder revealed more diverse interactions among nonmission research, mission-oriented research, and development activities.

Many insights revealed by TRACES have implications on today's discussions and policies concerning peer review and transformative research. Perhaps the most important lesson learned is the role of basic research or nonmission research. TRACES shows that an ultimate invention tends to result from the integration of multiple lines of research. Each line of research is often led by years and even decades of nonmission research, which is then in turn followed by mission-oriented research and development and application events. In other words, it is evident that it is unlikely for nonmission research to foresee how their work will evolve and that it is even harder for nonmission research in one subfield to recognize potential connections with critical development in other subfields. Bringing these factors together, we can start to appreciate the magnitude of the conceptual gulf that transformative research has to bridge.

2.2 CREATIVITY

Creativity can be seen as finding ways to break away from our current mental models. Researchers and practitioners have repeatedly asked whether creativity is something we were born with or something that can be trained and learned. The practical impli-cations are clearly related to the fact that individuals in organizations are expected to become increasingly creative as they collaborate in project teams. A meta-analysis conducted by Scott and colleagues (2004) reviewed the results of 70 studies of creative training effects and found that carefully constructed creativity training programs typically improve performance. In contrast, Benedek and colleagues (2006) studied whether repeated practice can enhance the creativity of adults in terms of the fluency and originality of idea generation. They found that while training did improve the fluency, no impact on originality was found.

The American psychologist Howard E. Gruber (1922–2005), a pioneer of the psychological study of creativity, questioned the validity of lab-based experimental studies of creativity. He argued that because creative works tend to be produced over a much longer period of time than the duration of a lab experiment, the laboratory setting is simply not realistic enough to study creativity. As a result, Gruber (1992) was convinced that an alternative approach, the evolving systems, should be used for the study of creativity. To him, a theory of creativity should explain the unique and unrepeatable aspects of creativity rather than the predictable and repeatable aspects seen in normal science.

Gruber strongly believed that the most meaningful study of creative work should focus on the greatest individuals rather than attempt to develop quantitative measures of creativity based on a larger group of people. His work, *Darwin on Man: A Psychological Study of Scientific Creativity*, is a classic exemplar of his evolving systems approach. His principle is similar to Albert Einstein's famous principle: as simple as it is, but not simpler. He strongly believed that characteristics of the truly creative work may not be found in an extended population (Gruber, 1992). Instead, he chose to study how exactly the greatest creative people such as Charles Darwin made their discoveries. He chose in-depth case studies over lab-based experimental studies.

From a sociological perspective of the philosophy of science, Randall Collins (1998) argued that intellectual life is first of all conflict and disagreement. His insight is that the advance of an intellectual field is very much due to rivalry and competing schools of thought that are often active within the same generational span of approximately 35 years. He introduced the notion of attention space and argued that "creativity is the friction of the attention space at the moments when the structural blocks are grinding against each other the hardest" (p. 76). The attention space is restructured by pressing in opposing directions. He spent over 25 years to assemble intellectual networks of social links among philosophers whose ideas have been passed along in later generations. He constructed such networks for China, India, Japan, Greece, modern Europe, and other areas over very long periods of time. He used a generation of philosophers as a minimal unit for structural change in an intellectual attention space. For example, it took six generations to move from Confucius to Mencius and Chuang Tzu along the Chinese intellectual chains. A major difference between Collins's grinding attention space and Kuhn's competing paradigms is that for Collins, explicit rivalry between schools of thought often developed in succeeding generations (Collins, 1998, p. 6), whereas Kuhn's competing paradigms are simultaneous. Major philosophers such as Mencius and Kung-Sun Lung are all at the center of colliding perspectives.

2.2.1   Divergent Thinking

Contrary Imaginations was a citation classic written by the British psychologist Liam Hudson (1966). It was identified as a citation classic in the October 1980 issue of *Current Contents*. Hudson noticed that schoolboys seem to have different levels of abilities to handle *convergent* and *divergent* questions. A typical convergent question gives multiple possible answers for an individual to choose, for example:

Brick is to house as plank is to

a. *Orange*
b. *Grass*
c. *Egg*
d. *Boat*
e. *Ostrich*

In contrast, a divergent question is an open-ended question that may have a numerous number of answers, like the question:

How many uses can you think of for a brick?

More interestingly, individuals differ considerably in terms of the number of answers they come up with. Some can think of many different ways to use a brick, whereas others may be only able to think of one or two. Based on the ability of an individual to answer these types of questions, Hudson differentiated individuals in terms of two intellectual types: *convergers*, who would specialize in mathematics and physical sciences, and *divergers*, who are likely to excel in the arts and make surprising cognitive leaps.

The 1981 Nobel Prize in Medicine was awarded to Roger Sperry for his pioneering work on split brain, which revealed the differences between hemispheres of the cerebral cortex and how the two hemispheres interact. The two hemispheres of the brain are connected. Each hemisphere has different functions. It is essential for the two hemispheres to communicate and function as a whole. If the connection between the two hemispheres is damaged, it results in a so-called split brain. Studies of split-brain patients show that the left brain is responsible for analytic, logical, and abstract thinking such as speaking, writing, and other verbal tasks, whereas the right brain is responsible for intuitive and holistic thinking such as spatial and nonverbal tasks. The right brain is also believed to home divergent thinking.

Divergent thinking has been widely regarded as a major hallmark of creativity. The distinction between convergent and divergent thinking was first made by the American psychologist Joy Paul Guilford (1897–1987), one of the pioneers in the psychometric study of creativity. He suggested that a prime component of creativity is divergent thinking (Guilford, 1967). The original terms were convergent and divergent production.

According to Guilford, divergent thinking is characterized by the presence of four types of cognitive ability (Guilford, 1967):

1. *Fluency*—the ability to produce a large number of ideas or solutions to a problem rapidly
2. *Flexibility*—the ability to consider a variety of approaches to a problem simultaneously
3. *Originality*—the tendency to produce ideas different from those of most other people
4. *Elaboration*—the ability to think through the details of an idea and carry it out

In contrast, convergent thinking is the ability to converge all possible alternatives to a single solution. When we take a test with multiple choice questions, we are typically using convergent thinking.

Following Guilford, divergent thinking is essential to creativity because it brings together ideas across different perspectives to reach a deeper understanding of a phenomenon. Divergent thinking can be easily recognized by its considerable variety and originality of the ideas generated. A large number of tests have been developed to measure the capacity for divergent thinking. Tests like *Alternate Uses* ask individuals to come up with as many different ways of using a common object as possible, such as a paper clip or a brick. Nevertheless, research has shown that one should be cautious in interpreting the implications of such measurements because this type of measurement may not adequately take the context of creativity into account. It is essential that creativity be studied in an appropriate context. After all, the ability of divergent thinking with a paper clip may tell us little about an individual's talent in music.

How should we deal with divergent thinking and convergent thinking in scientific discovery? Kuhn presented his view on this issue in a chapter of *The Essential Tension* (Kuhn, 1977), which was based on his 1959 speech at a conference on the identification of scientific talent. While recognizing the predominant attention to divergent thinking in scientific discovery, Kuhn argued that one should really take convergent thinking as well as divergent thinking into account simultaneously because it is the dynamic interplay between the two forms of thinking that drives the scientific creativity. A common misconception of Kuhnian scientific revolutions is that they are rare and separated by prolonged uneventful normal science. The misconception still exists today despite the fact that Kuhn pointed out the misconception early on. In addition to groundbreaking and world-shaking scientific revolutions, scientists experience numerous conceptual revolutions at much smaller scales. Because divergent thinking plays a dominant role in initiating a change of world views and convergent thinking plays a crucial role in consolidating the new direction, the tension between the two forms of thinking is essential. Just as that divergent thinking is *yin* and convergent thinking is *yang*, the development of science is a holistic process of antithesis elements! This type of interplay can be seen in the following Sections.

2.2.2 Blind Variation and Selective Retention

Many researchers have been deeply intrigued by the analogy between trial-and-error problem solving and natural selection in evolution. Is it merely an analogy on the surface or more than that? The American social scientist Donald Campbell (1916–1996) was a pioneer of one of the most profound creative process models. He characterized creative thinking as a process of *blind variation and selective retention* (Campbell, 1960). His later work along this direction became known as a *selectionist* theory of human creativity. If divergent thinking is what it takes for blind variation, then convergent thinking certainly has a part to play for selective retention.

Campbell was deeply influenced by the work of Charles Darwin. In Campbell's *evolutionary epistemology*, a discoverer searches for candidate solutions with no prior knowledge of whether a particular candidate is the ultimate one to retain. Campbell specifically chose the word *blind*, instead of *random*, to emphasize the absent of foresight in the production of variations. He argued that the inductive gains in creative processes hinge on three necessary conditions:

1. There must be a mechanism for introducing variation.
2. There must be a consistent selection process.
3. There must be a mechanism for preserving and reproducing the selected variations.

Campbell's work is widely cited, including both supports and criticisms. As of July 2010, his original paper was cited 373 times in the Web of Science and over 1000 times on Google Scholar.

Dean Simonton's *Origins of Genius: Darwinian Perspectives on Creativity* (1999) is perhaps the most prominent extension of Campbell's work. His main thesis was that the Darwinian model might actually subsume all other theories of creativity as special cases of a larger evolutionary framework. Simonton pointed out that there are two forms of Darwinism. The primary form concerns biological evolution. This is the Darwinism that is most widely known. The secondary form of Darwinian provides a generic model that could be applied to all developmental or historical processes of blind variation and selective retention. Campbell's evolutionary epistemology belongs to the secondary form of Darwinism. Campbell's proponents argued that the cultural history of scientific knowledge is governed by the same principles that guide the natural history of biological adaptations. Simonton provided supportive evidence from three methodological domains: the experimental, the psychometric, and the historiometric domains.

Critics of Campbell's (1960) model mostly questioned the blind variation aspect of the model. A common objection is that there would be too many possible variations to search if there is no effective way to narrow down and prioritize the search. The searcher would be overwhelmed by the enormous volume of potential variations. In contrast, the number of variations that would be worth retaining is extremely small. A scenario behind the *British Museum Algorithm* may illustrate the odds

(Newell, Shaw, & Simon, 1958). Given enough time, what are the odds of a group of trained chimpanzees typing randomly and producing all of the books in the British Museum?

Campbell defended his approach with the following key points and argued that the disagreement between his approach and the creative thinking processes of Newell, Shaw, and Simon was minor matters of emphasis.

1. There is no guarantee of omniscience. Not all problems are solved. Not all excellent solutions archived.
2. One should not underestimate the tremendous number of thought trails that did not lead to anywhere. What has been published in the literature is a small proportion of the total effort of the entire intellectual community. What has been cited is an even smaller proportion of it.
3. Selective criteria imposed at every step greatly reduce the number of variations explored.

The course of problem solving may drift away from the original goal and lead to unexpected achievements such as solving a new problem. Campbell positioned his work as a perspective on creative thought processes rather than a theory of creative thinking. A perspective merely points to the problems, whereas a theory specifies underlying mechanisms and makes predictions. Nevertheless, one should ask whether creative thinking processes are predictable by their very nature. Recall Kuhn's dialectic view of the essential tension between divergent and convergent thinking: divergent thinking is crucial in the blind variation stage of the process, whereas convergent thinking is important in the selective retention stage.

Other critics of Campbell's work argue that his model is not broad enough to account for the full spectrum of creativity. Campbell's supporters defended that although the process may not be involved in all forms of human behavior and thought, Campbell had made a compelling case that all genuine forms of human creativity and invention are characterized by the process (Cziko, 1998).

2.2.3   Binding Free-Floating Elements of Knowledge

The blind variation and selective retention perspective is also evident in the work of the late Chinese scholar Hongzhou Zhao, although it is not clear whether his work was influenced by Campbell's work. Outside China, Zhao is probably better known for his work on the dynamics of the world center of scientific activities (Zhao & Jiang, 1985). With his education in physics, Zhao defined an element of knowledge as a scientific concept with a quantifiable value, for instance, the concepts of force and acceleration in Newton's $F = ma$, or the concept of energy in Einstein's $E = mc^2$. The mechanism for variation in scientific discovery is the creation of a meaningful binding between previously unconnected knowledge elements in a vast imaginary space. The complexity of an equation can be quantified in terms of the "weight" of the elements involved.

The space of knowledge elements contains all the knowledge that has ever been created. The number of potentially valuable elements is huge. However, only a small fraction of all these bindings are potentially meaningful and desirable. There were a total of N elements of mechanics just before Newton discovered his second law. The number of candidate elements for binding is the number of ways to choose the three elements F, m, and a. If there were Q ways to connect these elements with operations such as addition and subtraction, then Newton could expect to find his answer in $W = A_N^3 A_Q^1$ possible variations. In general, the number of paths to explore would be $W = A_N^e A_Q^{e-2}$, where e is the number of elements involved and Q is the number of operations. Furthermore, one can define the entropy of knowledge S as $\ln\left(A_N^e A_Q^{e-2}\right)$. Creative thinking is to reduce the entropy S by binding certain elements.

The entropy S has an alternative interpretation—it measures the potential of creation. The number of candidate elements in our knowledge varies with age and expertise. To acquire a new knowledge element, we need to have a thorough understanding of a concept. If an element is available for a possible binding with other elements, it is called a free-floating element. A beginner would have a rather small number of such elements to work with, so the potential of their creativity is low. In contrast, established scientists would have a large amount of knowledge elements, but they are likely to be biased from their existing mental models and often fail to see new options for binding. They tend to interpret new things in terms of an old perspective or theory. As a result, the number of truly available free-floating knowledge elements is also quite limited for these established scientists. Zhao used this framework to argue for the existence of an optimal age when the highest entropy S is reached. At the optimal age, the scientist would have not only enough free-floating knowledge elements to work with but also the least amount of bias that could hinder their ability to see new ways of binding knowledge elements.

The earlier description of Zhao's model identifies situations where the number of meaningful connections is overwhelmingly outnumbered by the size of a vast space of potentially viable candidates. Examples include the study of the origin of the universe in astronomy, searching for biologically and chemically constrained compounds in the vast chemical space in drug discovery and searching for new connections between previously disparate bodies of knowledge in literature-based discovery. Campbell was right—searching for a small number of meaningful solutions in a boundlessly large space seems to be a common problem. Indeed, Zhao's element-binding model is almost identical to Campbell's theory of creative discoveries in science, that is, a blind variation and selective retention process.

Zhao was searching for insights that could further explain what would make the center of scientific activities shift. For instance, how would the optimal age of a scientist influence the quality of element-binding activities? The center of scientific activity refers to the country that has more than 25% of noteworthy contributions of the entire world. The notion of the optimal age suggests that the majority of scientists at their optimal age are most creative. From the knowledge element-binding perspective, scientists who are younger than the optimal age may have the best memory, but they are probably inexperienced and have not seen enough examples. Therefore, their ability to see potential and alternative ways of binding elements of knowledge would not be as efficient as those who are in their optimal ages. Zhao found that when a country's social mean age and the optimal age distribution get close to each other, the country's science is likely on the rise. In contrast, when the social mean age drifts away from the optimal age distribution, the country's science is likely in decline. If a country is to become the center of scientific activity, its scientists must be in the best shape to bind knowledge elements in the most effective and creative way.

The strength of Zhao's knowledge element approach is that it can be used to identify a macroscopic movement of scientific activity worldwide. It focuses on the high-level description of a generic binding mechanism. It has practical implications for science policy of a country. Unfortunately, Zhao's approach has shortcomings shared by many statistical approaches. It does not lend itself as a concrete step-by-step procedure that an individual scientist can pursue in daily activities. It does not give guidelines on how a specific binding should be conceived. The mechanism for making new variations is completely blind. Each variation does not make the next variation any easier. We have a problem!

2.2.4 Janusian Thinking

Janusian thinking is a special form of divergent thinking. Shakespeare's Hamlet was torn between two conflicting and mutually exclusive choices that he had to make: either *to be* OR *not to be*. In contrast, the idea of Janusian thinking is to find a new perspective that can satisfy both *to be* AND *not to be* simultaneously.

Janusian thinking is proposed by Albert Rothenberg (1996) in 1979 as a process that aims at *actively conceiving* multiple opposites or antitheses simultaneously. It is named after the Roman god Janus, who had two faces looking in opposite directions. As we shall see shortly, Janusian thinking can be seen as a special type of divergent thinking, and it can be used to generate original ideas systematically. In addition, an interesting connection between Janusian thinking and the work of the sociologist Murray S. Davis on why we find a new theory interesting is discussed in Chapter 4.

Rothenberg studied Nobel Prize winners, creative scientists, and artists. He came to realize that this is the type of thinking used by Einstein, Bohr, Mozart, Picasso, and many others in their creative work. For instance, the notion of symmetry has served the role of a variation mechanism in many famous scientific and mathematical discoveries—such as Einstein's relativity and Bohr's complementarity. Rothenberg was convinced that most major scientific breakthroughs and artistic masterpieces are resulted from Janusian thinking.

Janusian thinking progresses through four phases over extended periods of time: (1) motivation to create, (2) deviation or separations, (3) simultaneous opposition or antithesis, and (4) construction of the theory, discovery, or experiment. Typically, Janusian thinking starts by asking: What is the opposite of a concept, an interpretation, or a phenomenon? In next phase, scientists start to break away from the work of other scientists. The deviation does not usually occur all at once but as an evolution from previous thought. In phase 3, opposites are juxtaposed simultaneously in a conception, and this conception is then transformed in phase 4 and leads directly to the creative outcome. Phase 4 represents the construction of the full dimensions of the theory or discovery. Ultimately, the creative scientists encapsulate an area of knowledge and relations when bringing and using these opposites *together*. The nature of the new conception is similar to a Gestalt switch or a change of viewpoint. Finding the right perspective is the key to creativity.

It is common to see that even after the critical conception is in place, there is usually still much of work to be done (Rothenberg, 1996). The construction phase includes components such as articulation, modifications, and transformations to produce integrated theories and discoveries. Articulation keeps elements separated but connected. In this sense, Janusian thinking is a brokerage or boundary-spanning mechanism that connects opposites and antitheses. We will discuss the brokerage-based theory of discovery in more detail in the following chapters. The brokerage theory of discovery was originally proposed in Chen et al. (2009).

Researchers have noticed that scientists who made great discoveries may experience solutions and discoveries before their ultimate discoveries. For Darwin, it all came together when he was reading Thomas Malthus's *Population* (1826). Malthus's main point was that overpopulation of a species in a closed environment would eventually lead to a devastating destruction of the species due to competition for existence. Darwin had read Malthus several times, and he could see how the very same struggle for existence increasingly enhanced the odds of survival of the species in its environment. Favorable and unfavorable variations of species have to be considered *at once* and formed a simultaneous antithesis. Darwin depicted each and every known species as separated and connected with one another through evolution. Malthus's work also inspired A. R. Wallace when he independently conceived an idea of natural selection. The ability to see how oppositions, favorable–unfavorable and increase–decrease, work together simultaneously turned out to be critical in both of the creative discoveries of the principle of natural selection.

The trail of Janusian thinking was also evident in Danish physicist Bohr's discovery of the complementarity principle. Bohr's thinking experienced a shift from the principle of correspondence to a breakthrough conception of simultaneous antithesis. The complementarity principle explained how light could be both wave and particle simultaneously. Two mutually exclusive appearances are necessary for a comprehensive description of the situation. Bohr discovered that the two concepts of wave and particle were a description of two complementary sides of a single phenomenon—light.

Rothenberg (1996) interviewed 22 Nobel laureates in the fields of chemistry, physics, and medicine and physiology from Europe and the United States. The interviews followed a systematic research protocol focused on in-progress creative work. *Einstein, Bohr, and Creative-Thinking in Science* (Rothenberg, 1987) included analyses of autobiographical accounts and work-in-progress manuscripts pertaining to the creative formulations and discoveries of outstanding scientists of the past, such as Bohr, Darwin, Dirac, Einstein, Planck, and Yukawa. Einstein's general theory of relativity was used as an example to show the nature of his first thought as an antithetical idea. As Einstein recalled that the idea was "for an observer in free fall from the roof of a house, there exists, during his fall, no gravitational field … in his immediate vicinity. If the observer releases any objects, they will remain, relative to him, in a state of rest." The question is: how can gravity be seen as present and absent simultaneously?

All of the scientists interviewed by Rothenberg specifically spoke of the advantage of coming fresh into a new area and not being preoccupied by some of the biases and assumptions of that area. Breaking away from existing perspectives is critical in creative thinking. One of the commonly used mechanisms of creative imagination is to use distant analogies to escape the constraints of our existing mental models, especially at early stages of the creative process. Research has shown that scientists at the initial stage of a revolutionary discovery often use distant analogies, and as they develop a better grasp of the nature of the problem, these distant analogies are gradually replaced with analogies that are near to the problem at hand. In Chapter 4, we will discuss another relevant phenomenon called Proteus phenomenon.

Rothenberg noted that creative people may not always realize that they are taking these steps in their thinking themselves, but the trails of these steps can be traced retrospectively. Rothenberg argued that Janusian thinking works because oppositions and antitheses represent polarities and extremes of a scale or category. Their explicit involvement in the discovery process provides a basis for storing and extending knowledge (Rothenberg, 1996, p. 222). In other words, the extreme instances function just as landmarks for scientists finding their paths and impose constraints on potentially valuable variations. These landmarks are valuable because they clarify the nature and content of intermediary factors and, more importantly, the latent and previously unknown category.

From a philosophical point of view, Murray Davis proposed an intriguing framework to explain why we find some theories more interesting than others (Davis, 1971). In essence, a new theory would be regarded as interesting if it triggers a switch of our perspectives from one that we have taken for granted to a view that may contradict to what we believe. The caveat is that the new theory should not overdo it. If it goes too far, it will lose our interest. The difference between Davis's framework and Janusian thinking is subtle but significant. In Davis's framework, when we are facing two opposite and contradictory views, we are supposed to choose one of them. In contrast, Janusian thinking is not about choosing one of the existing views and discarding the other. Instead, we must come up with a new and creative perspective so that it can accommodate and subsume all the contradictions. The contradictions at one level are no longer seen as a problem at the new level of thinking. It is in this type of conceptual and cognitive transformation that discoverers create a new theory that makes the coexistence of the antitheses meaningful.

The ability to view things from multiple perspectives and reconcile contradictions is in the center of dialectical thinking. The origin of dialectics is a dialogue between two or more people with different views but wish to seek a resolution. Socratics, Hegel, and Marx are the most influential figures in the development of dialectical thinking.

According to Hegel, a dialectic process consists of three stages of thesis, antithesis, and synthesis. An antithesis contradicts and negates the thesis. The tension between the thesis and antithesis is resolved by synthesis. Each stage of the dialectic thinking process makes implicit contradictions in the preceding stage explicit. An important dialectical principle in Hegel's system is the transition from quantity to quality. In the commonly used expression "the last straw that broke the camel's back," the one additional straw is a quantitative change, where a breakdown camel is a qualitative change. The negation of the negation is another important principle for Hegel. To Hegel, human history is a dialectical process.

Hegel was criticized by materialist or Marxist dialectics. In Karl Marx's own words, his dialectic method is the direct opposite of Hegel's. To Marx, the material world determines the mind. Marxists see contradiction as the source of development. In this view, class struggle is the contradiction that plays the central role in social and political life. In Chapter 1, we introduced how internalism and externalism differ in terms of their views of the nature of science and its role in the society. Dialectic thinking does seem to have a unique place in a diverse range of contexts.

Opposites, antitheses, and contradictions in Janusian thinking in particular and dialectic thinking in general are integral part of a broader system or a longer process. Contradictory components are not reconciled but remain in conflict; opposites are not combined, and oppositions are not resolved (Rothenberg, 1996). Opposites do not vanish; instead, one must transcend the tension between contradictory components to find a creative solution.

With regard to the five-stage model of a creative process, the most creative and critical components of Janusian thinking are the transition from the third phase to the fourth phase, that is, from simultaneous opposition to the construction of a new perspective. In Campbell's perspective of blind variation and selective retention, Janusian thinking proactively seeks antitheses as a mechanism for variation and imposes retention criteria for variations that can synthesize theses and antitheses.

2.2.5 TRIZ

TRIZ is a method for solving problems innovatively. The Russian acronym TRIZ is translated into English as the Theory of Inventive Problem Solving (TIPS). It was originally developed by the former Soviet engineer and researcher Genrich Altshuller (1926–1998). His earlier experience at the Naval Patent Office was believed to influence his development of TRIZ. The landmark work on TRIZ is *The Innovation Algorithm* (Altshuller, 1999). It was first published in 1969. Its English translation was published in 1999.

The Innovation Algorithm has three sections: technology of creativity, dialectics of invention, and man and algorithm. Altshuller analyzed different methods of technical creativity, and he was convinced that people can be trained to be innovative. He developed 40 principles that can be used to resolve technical contradictions. To Altshuller, the main obstacles to creativity are psychological barriers. These obstacles can be overcome through a higher creative consciousness—a TRIZ mind.

The entire TRIZ method is built on the belief that the process of creativity can be learned (Altshuller, 1999). The process of creativity can be detected and made accessible to those who need to solve problems creatively. An "algorithm" or a recipe is available for invention. Altshuller described how an inventor follows a path of a trial-and-error search: "Eventually, an idea emerges: 'What if we do it like this?' Then, theoretical and practical testing of the idea follows. Each unsuccessful idea is replaced with another, and so on." This is indeed, as you have recognized it, the same idea as Campbell's blind variation and selective retention.

Altshuller realized that an inventor may have to face a severe challenge. The inventor would have to search in a potentially very large space of possible solutions. There are much fewer valid solutions than nonusable solutions. The core insight from Altshuller is summarized as a set of 40 principles that one can follow in searching for paths to inventions. These principles form a systematic approach to the invention of new solutions and the refinement of existing solutions. It is designed to facilitate inventors to formulate problems in a way that is strikingly similar to Janusian thinking. In a nutshell, TRIZ is a constructive approach to creative thinking.

TRIZ focuses on technical contradictions and how to remove them. From this perspective, an invention is the removal of technical contradictions. Here are some examples of the 40 principles given in *the Innovation Algorithm*:

Segmentation
- Divide an object into independent parts.
- Make an object sectional (for easy assembly or disassembly).
- Increase the degree of an object's segmentation.

Extraction
- Extracting, retrieving, and removing.
- Extract the "disturbing" part or property from an object.
- Extract only the necessary part or property from an object.

Do It in Reverse
- Instead of the direct action dictated by a problem, implement an opposite action (i.e., cooling instead of heating).
- Make the movable part of an object, or outside environment, stationary—and stationary part movable.
- Turn an object upside down.

Convert Harm to Benefit
- Utilize harmful factors—especially environmental—to obtain a positive effect.
- Remove one harmful factor by combining it with another harmful factor.
- Increase the degree of harmful action to such an extent that it ceases to be harmful.

The invention of light bulb illustrates how TRIZ works.[1] It was known as early as 1801 that when electric current passes through metal filaments, filaments will

---

[1]http://www.salon.com/tech/feature/2000/06/29/altshuller/index.html

light up. But the problem was that the filaments burned out in the process. The contradiction is that the filaments must get hot enough to glow but not too hot to burn themselves out. The contradiction was not resolved until 70 years later by the invention of Joseph Wilson Swan and Thomas Alva Edison. They solved the problem by placing the filaments in a vacuum bulb.

Another classic example is the design of tokamak for magnetic confinement fusion. Tokamaks were invented in the 1950s by Soviet physicists Igor Tamm and Andrei Sakharov. The problem was how to confine the hot fusion fuel plasma. The temperature of such plasmas is so high that containers made of any solid material would melt away. The contradiction was resolved by using a magnetic field to confine the plasma in the shape of a doughnut. Contradiction removal is a valuable strategy for creative thinking and problem solving in general. Focusing on a contradiction is likely to help us to ask the right questions.

2.2.6   Reasoning by Analogy

Aliens in Hollywood movies are often humanlike creatures, computer-transformed earth animals, or combinations of both. When college students were asked to create imaginary creatures, they would expect to encounter on an alien planet, most of them recombine and reorganize features found on earth animals. Our imagination is fundamentally biased by what we have seen and what we have experienced.

Reasoning by analogy is a frequently used strategy in science. The strategy duplicates the path of a previous success. One example is found in mass extinction research. An impact theory was originally proposed in 1980 to explain what caused the mass extinctions about 65 million years ago, known as the KT mass extinction. The extinction of dinosaurs was one of the consequences of the mass extinction. The impact theory suggested that an asteroid collided into the Earth and the dust of the impact covered the Earth's atmosphere for a couple of years. In 1990, a big crater was discovered in the Mexico Bay, and the crater was believed to be the most direct evidence for the impact theory. In 2001, inspired by the success of the impact theory in explaining the KT extinction, researchers proposed a new line of research that would follow the same pattern of the impact theory's success. The difference was that it aimed to explain an even earlier mass extinction 250 million years ago. However, the validity of the analogy was questioned by more and more researchers. By 2010, it is the consensus that the analogy does not seem to hold given the available evidence. We were able to detect the analogical path from citation patterns in the relevant literature in our paper published in 2006 (Chen, 2006). The same conclusion was reached by domain experts in a 2010 review (French & Koeberl, 2010). We will revisit this example in later chapters of this book.

2.2.7   Structural Holes, Brokerage, and Boundary Spanning

Structural holes are defined as a topological property of a social network. According to Burt (2004), structural holes refer to the lack of comprehensive connectivity among components in a social network. The connection between distinct components is reduced to few person-to-person links at structural holes, whereas person-to-person links tend to be uniformly strong at the center of a group. Because information flows are restricted to the privileged few who are strategically positioned over structural holes, the presence of a structural hole has a potential for gaining distinct advantages.

Opportunities generated by structural holes are a vision advantage. People connected across groups are more familiar with alternative ways of thinking, which gives them more options to select from and synthesize. Because they have more alternative ideas to choose from, the quality of the selection tends to be better.

Burt identifies four levels of brokerage through which one can create value, from the simplest to the most advanced:

1. Increasing the mutual awareness of interests and difficulties of people on both sides of a structural hole
2. Transferring best practice between two groups
3. Drawing analogies between groups seemingly irrelevant to one another
4. Synthesizing thinking and practices from both groups

Burt found indeed that a vision advantage associated with brokerage translates to better received ideas.

A brokerage between information foraging and the structural hole theory may be fruitful for visual analytics. The structural-hole-induced brokerage perspective addresses situations where information scents are either missing or unreachable. On the other hand, the structural hole theory can guide the selection of potentially information-rich paths for foragers. A brokerage-oriented perspective focuses on the linkages connecting distinct components in a complex system. Consequently, the focus on intercomponent connections provides us a unique leverage to differentiate the role of individual entities at a higher level of aggregation.

In the research literature of terrorism, for example, the earliest theme is about physical injuries; a later theme is centered on health care and emergency responses; the most recent theme focuses on psychological and psychiatric disorders. Cognitive transitions from one theme to another become easier to grasp at this level. This high-level understanding can also serve a meaningful context for us to detect what is common and what is surprise. To understand terrorism research as a whole, it is necessary to understand how these themes are interrelated. The whole here is indeed more than the sum of parts. An information-theoretic view brings us a macroscopic level of insights.

The role of spanning over a structural hole in a network underlines a theme that has been found in several distinct perspectives. Norbert Wiener wrote in his famous book on cybernetics that the most fruitful areas for the growth of the sciences were where had been neglected as a no man's land between various established fields. In terms of the landscape metaphor, which we will discuss in more detail in later chapters, established fields would correspond to peaks or mountains on a landscape representation of the level of activities, whereas a no man's land in between would be valleys or even uncharted waters that no activities had been recorded and no attention was ever directed toward them.

In Wiener's word, "A proper exploration of these blank spaces on the map of science could only be made by a team of scientists, each a specialist in his own field but each possessing a thoroughly sound and trained acquaintance with the fields of his neighbors." His description provides an insightful description of today's interdisciplinary research. The divisions of disciplines tend to outlive the original motivating problems that the expertise was gathered to address. A mismatch between a new problem and what is needed to address it would explain why the forefront of scientific research seems to be on a constant shift toward structural holes and no man's lands. Our current mental models become out of date and out of focus in light of new problems. We need to transform our mental models and refocus.

2.3 FORESIGHTS

2.3.1 Information Foraging

Information foraging theory is a predictive model of information foragers' search behavior (Pirolli, 2007). According to the theory, an information environment is made of patches of information, and an information forager moves from one patch to another as they search for information, just as a predator looking for its prey in the animal world. The theory is developed to answer questions about how information foragers would choose patches to work with and what influence their decisions on how they should spend their time with patches.

Information foraging theory is built on the assumption that people adapt their search strategies to maximize their profitability or the profit–investment ratio. People may adapt their search by reconfiguring the information environment. The investment typically includes the time spent in searching and assimilating information in patches of information. The profit includes the gain by finding relevant information. Users, or information foragers, tend to follow a path that can maximize the overall profitability. *Information scent* is the perception of the value, cost, or accessible path of information sources. When possible, one relies on information scent to estimate the potential profitability of a patch.

The power of information foraging theory is its own adaptability and extensibility. It provides a quantitative framework for interpreting behavioral patterns at both microscopic and macroscopic levels. For instance, connecting the dots of mysterious behaviors of 9/11 hijackers at flying schools would depend on the prevalence and strength of the relevant information scent (Anderson, Schum, & Twining, 2005). The question is where an analyst could draw the right information scent in the first place. Figure 2.3 is not designed with information foraging theory in mind, but the visualization intuitively illustrates the profit maximization principle behind the theory. The connective density reinforces the boundaries of patches. Colors and shapes give various information scents about each patch in terms of its average age and the

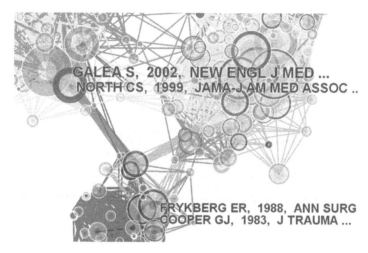

GALEA S, 2002, NEW ENGL J MED ...
NORTH CS, 1999, JAMA-J AM MED ASSOC ..

FRYKBERG ER, 1988, ANN SURG
COOPER GJ, 1983, J TRAUMA ...

FIGURE 2.3 The three clusters of cocited papers can be seen as three patches of information. All three patches are about terrorism research. Prominently labeled papers in each patch offer information scent of the patch. The sizes of citation rings provide a scent of citation popularity. Source: From Chen (2008).

popularity of citations. These scents will help users to choose which patch they want to explore in more detail.

From the information-theoretic view, information scent only makes sense if it is connected to the broader context of information foraging, including the goal of search, the prior knowledge of the analyst or the information forager, and the contextual situation. This implies a deeper connection between the information-theoretic view and various analytic tasks in sensemaking. The following is a sensemaking example in which an information-theoretic approach is applied to the study of uncertainty and influential factors involved in political elections.

Voting in political elections involves a complex sensemaking and reasoning process. Voters need to make sense overwhelmingly diverse information, differentiate political positions, accommodate conflicting views, adapt beliefs in light of new evidence, and make macroscopic decisions. Information-theoretic approaches provide a valuable and generic strategy for addressing these issues. Voters in political elections are influenced by candidates' positions regarding a spectrum of political issues and their own interpretations of candidates' positions (Gill, 2005).

Researchers are particularly interested in the impact of uncertainty concerning political positions of candidates from a voter's point of view. From an information-theoretic perspective, candidate positions on a variety of controversial issues can be represented as a probability distribution. The underlying true distribution is unknown. The voters' uncertainty can be measured by the *divergence* of a sample from the true distribution. In a study of the 1980 presidential election, Bartels finds that voters in general dislike uncertainty (Bartels, 1988).

In a study of a 1994 congressional election, Gill (2005) constructs an aggregate measure of uncertainty of candidates as well as political issues based on voters' self-reported three-level certainty information regarding each political issue question. The study analyzed answers from 1795 respondents on several currently salient issues, such as crime, government spending, and health care. The results suggest that politicians are better off with unambiguous positions, provided that those positions do not drastically differ from those held by widely supported candidates.

The effect of uncertainty seems to act at aggregated levels as well as individual levels. Crime, for example, has been a Republican campaign issue for decades. The study found that Republican candidates who are vague on this issue are almost certainly punished (Gill, 2005, p. 387). This example shows that an information-theoretic approach provides a flexible tool for studying information uncertainty involved in complex reasoning and decision-making processes.

2.3.2 Identifying Priorities

Goodwin and Wright (2010) reviewed forecasting methods that target for rare and high-impact events. They identified the following six types of problems that may undermine the performance of forecasting methods:

1. Sparse reference class.
2. Reference class is outdated or does not contain extreme events.
3. Use of inappropriate statistical models.
4. The danger of misplaced causality.
5. Cognitive biases.
6. Frame blindness.

Goodwin and Wright identified three heuristics that can lead to systematically biased judgments: (1) availability, (2) representativeness, and (3) anchoring and insufficient adjustment. The availability heuristic bias means that human beings find easier to recall some events than others, but it usually does not mean that easy-to-recall events have a higher probability of occurring than hard-to-recall events. The representativeness heuristic is a tendency to ignore base-rate frequencies. The anchoring and insufficient adjustment means that forecasters make insufficient adjustment for the future conditions because they anchor on the current value. As we can see, cognitive biases illustrate how vulnerable human cognition is in terms of estimating probabilities intuitively. Expert judgment is likely to be influenced by these cognitive biases. Researchers have argued that in many real-world tasks, apparent expertise may have little relationship to any real judgment skill at the task in question.

The increasing emphasis on accountability for science and science policy is influenced by many factors, but two of them are particularly influential and persistent despite the fact that they started to emerge more than a decade ago. The two factors are (1) limited public funding and (2) the growing view that publicly funded research should contribute to the needs of society (MacLean, Anderson, & Martin, 1998). The notion of a value-added chain is useful for explaining the implications. The earlier value-added chain, especially between the 1940s and 1960s, was simple. Researchers and end users were loosely coupled in such value-added chains. The primary role of researchers was seen as producing knowledge, and the primary role of end users was to make use of produced knowledge whenever applicable, passively. Decisions on how to allocate research funds were largely made based on the outcome of peer reviews. The peer here was the peer of scientists, and the peer of end users was not part of the game. The focus was clearly and often exclusively on science.

The view that science should serve the needs of society implies that the linkage between science and end users becomes an integral part of science policy and strategic long-term planning, including identifying funding priorities and assessing the impact of research. As a result, the new value-added chain includes intermediate users as well as scientists and end users. For example, following (MacLean et al., 1998), a simple value-added chain may include researchers in atmospheric chemistry, intermediate users from meteorological office and consultancy firms, and a supermarket chain as an end user. Several ways are suggested to understand users, including their long-term or short-term needs, generic versus specific needs, proactive compared to reactive users, and end users versus intermediate users. The role of intermediate users is to transform the scientific and technological knowledge and add value to such knowledge for the benefit of the following user in the chain. In one example of a complex value-added chain given in MacLean et al., the value-added chain consists of three categories of stakeholders: researchers, intermediate users, and end users. Aquatic pollution researchers may communicate with intermediate users such as sensor development companies, informatics companies, and pollution regulatory authorities. Intermediate users may have their own communication channels among themselves. Intermediate users in turn connect to water companies and polluting industry.

In soliciting users' opinions, especially from scientists, users tend to concentrate on shorter-term issues and more immediate problems than a 10- to 20-year strategic time frame. One way to encourage users to articulate their longer-term research needs is to ask users a set of open-ended questions. Here are some examples of open-ended questions on long-term environmental research priority issues (MacLean et al., 1998):

- *If someone can tell you precisely how environmental issues would affect your business in 10–20 years, what questions would you most wish to ask?*
- *What understanding about the environment do you not have at present, but would need in the next 5–20 years to enhance your organization's business prospects?*
- *If all the constraints, financial or otherwise, can be removed, what would you suggest that funding agencies could do in relation to environmental research in addition to what you have already mentioned?*

In assessing science and technology foresights, one way to identify priorities is to solicit assessments from experts and users and organize their assessments along two dimensions: feasibility and attractiveness. The dialogue between researchers and users is increasingly regarded as a necessary and effective approach to identify science and technology priorities. Scientists and researchers are likely to provide sound judgment on what is feasible, whereas users often make valuable input on what is attractive. This type of method was adopted by MacLean et al. (1998) to identify that the nature of links in a value-added chain is to map science outputs on to user needs. A two-round Delphi survey was conducted. Responses from over 100 individuals were obtained in each round. The differences between responses in the two rounds were plotted in a two-dimensional feasibility by attractiveness space.

The movements of assessments in this two-dimensional space represent the linkage between scientists and users in the value-added chain model. For example, the topic *remote data acquisition* in the high feasibility and high attractiveness moved to a position with an even higher feasibility and a higher attractiveness after the second round of the Delphi survey. In contrast, the topic *sustainable use of marine resources* was reduced in terms of both feasibility and attractiveness. It is possible that one group of stakeholders changed their assessments, but the other group's assessments remained unchanged. For example, while users did not alter their attractiveness assessments of topics such as management of freshwater resources and prediction of extreme atmospheric events, scientists updated the corresponding feasibility assessments: one went up and the other went down.

In contrast to the broadened social contract view of science in today's society, it is worth noting that there is a profound belief in the value of intellectual freedom and the serendipitous nature of science. Both sides have many tough questions to answer. Is there sufficient evidence based on longitudinal, retrospective, and comparative assessments of both priority areas chosen by science and technology foresight approaches and scientific and technological breakthroughs emerged and materialized regardless? Given the evidently changing consensus between different voting rounds of Delphi surveys, what are the factors that trigger the shift?

2.3.3 Hindsight on Foresight

How realistic and reliable are expert opinions obtained from foresight activities? There is a rich body of literature on Delphi and related issues such as individual opinion change and judgmental accuracy in Delphi-like groups (Rowe, Wright, & McColl, 2005), pitfalls, and neglected aspects (Geels & Smit, 2000).

Felix Brandes (2009) addressed this issue and assessed expert anticipations in the UK technology foresight program. The United Kingdom's technology foresight program was recommended in the famous 1993 White Paper "Realising our Potential." Fifteen expert panels were formed along with a large-scale national Delphi survey. The survey was sent to 8384 experts in 1994 to generate forecasts on 2015 and beyond for which 2585 responded to the survey. About two-thirds of statements were predicted to be realized between 1995 and 2004. Brandes's study was therefore to assess how realistic the 1994 expert estimates were by 2004, that is, 10 years later.

Out of the original 15 panels of the 1994 UK foresight program, Brandes selected three panels: Chemicals, Energy, and Retail & Distribution, to follow up the status of their forecasts in terms of *realized, partially realized, not realized*, and *don't know*. An online survey was used and the overall response rate was 38%.

Brandes's *Hindsight on Foresight*[2] survey found only 5% of the *Chemicals* statements, and 6% of the *Retail & Distribution* statements were regarded as realized by the experts surveyed, whereas 15% of *Energy* topics were realized. If the assessment criteria were relaxed to lump together fully and partially realized topics, known as the joint realization rate, Chemicals scored 28%, Energy 34%, and Retail & Distribution 43%.

In summary, the 1994 expert estimates were overly optimistic, which is a well-documented issue in the literature. Researchers have found that top experts tend to be even more optimistic than the overall response group (Tichy, 2004). According to Tichy the assessment of top experts tends to suffer from an optimism bias due to the experts' involvement and their underestimation of realization and diffusion problems. Experts with top working in business have a stronger optimism bias than those working in the academia or in the administration. Reasons that cause such biases are still not completely clear. More importantly, retrospective assessments of the status of priority topics identified by foresight activities are limited in that they do not provide overall assessments of topics and breakthroughs that were totally missed and unanticipated by foresight activities.

2.4 SUMMARY

Our judgments, decisions, and interpretations of information are driven by our mental models. A good mental model can guide us to discover otherwise unexpected details, whereas a wrong mental model may cost life. Mental models are easy to form. One may construct a mental model with a glimpse of the tip of an iceberg. Mental models have potentially severe downsides. Mental models are resilient to change. It is easy to get trapped in a mental model even when newly available information may indicate inconsistencies with the mental model. When newly available information appears to be inconsistent with our mental model, we are more likely to explain away the inconsistency than question the validity of the model. Explaining away inconsistent evidence is dangerous and risky because such explanations are often taken for granted with little or no verification. It is essential to be mindful when we should question the very model that is guiding our thinking.

Creativity is vital in breaking away from an established mental model. When we direct our attention to the validity of an established mental model, there is a chance that a breakthrough is underway. In successful stories such as the gene silencing discovery, an existing mental model gives way to more advanced models, which are subject to further tests and probing.

The information foraging provides an insightful understanding of strategies that often influence our decision making in general. When we consider whether we should challenge an established theory, we are thinking like a forager.

[2] The Office of Science and Technology (United Kingdom) sent out a "Hindsight on Foresight" survey in 1995.

BIBLIOGRAPHY

Altshuller, G. (1999). Innovation Algorithm: TRIZ, systematic innovation and technical creativity (1st ed.). Worcester, MA: Technical Innovation Center, Inc.

Anderson, T., Schum, D., & Twining, W. (2005). *Analysis of evidence* (2nd ed.). Cambridge, UK: Cambridge University Press.

Bartels, L. (1988). Issue voting under uncertainty: An empirical test. *American Journal of Political Science, 30,* 709–728.

Beadle, G. W. (1974). Recollections. *Annual Review of Genetics, 43,* 1–13.

Benedek, M., Fink, A., & Neubauer, A. C. (2006). Enhancement of ideational fluency by means of computer-based training. *Creativity Research Journal, 18,* 317–328.

Ben-Zvi, A. (1976). Hindsight and foresight: A conceptual framework for the analysis of surprise attacks. *World Politics, 28*(3), 381–395.

Betts, R. K. (2007). Two faces of intelligence failure: September 11 and Iraq's Missing WMD. *Political Science Quarterly, 122*(4), 585–606.

Bowker, G. C., & Star, S. L. (2000). *Sorting things out—Classification and its consequences.* Cambridge, MA: MIT Press.

Brandes, F. (2009). The UK technology foresight programme: An assessment of expert estimates. *Technological Forecasting and Social Change, 76*(7), 869–879.

Burt, R. S. (2004). Structural holes and good ideas. *American Journal of Sociology, 110*(2), 349–399.

Campbell, D. T. (1960). Blind variation and selective retentions in creative thought as in other knowledge processes. *Psychological Review, 67*(6), 380–400.

Chassin, M. R., & Becher, E. C. (2002). The wrong patient. *Annals of Internal Medicine, 136,* 826–833.

Chen, C. (2006). CiteSpace II: Detecting and visualizing emerging trends and transient patterns in scientific literature. *Journal of the American Society for Information Science and Technology, 57*(3), 359–377.

Chen, C. (2008). An information-theoretic view of visual analytics. *IEEE Computer Graphics & Applications, 28*(1), 18–23.

Chen, C., Chen, Y., Horowitz, M., Hou, H., Liu, Z., & Pellegrino, D. (2009). Towards an explanatory and computational theory of scientific discovery. *Journal of Informetrics, 3*(3), 191–209.

Chubin, D. E., & Hackett, E. J. (1990). *Paperless science: Peer review and U.S. science policy.* Albany, NY: State University of New York Press.

Collins, R. (1998). *The sociology of philosophies: A global theory of intellectual change.* Cambridge, MA: Harvard University Press.

Comroe, J. H., & Dripps, R. D. (2002). Scientific basis for the support of biomedical science. In R. E. Bulger, E. Heitman, & S. J. Reiser (Eds.), *The ethical dimensions of the biological and health sciences* (2nd ed., pp. 327–340). Cambridge, UK: Cambridge University Press.

Cziko, G. A. (1998). From blind to creative: In defense of Donald Campbell's selectionist theory of human creativity. *Journal of Creative Behavior, 32*(3), 192–209.

Davis, M. S. (1971). That's Interesting! Towards a phenomenology of sociology and a sociology of phenomenology. *Philosophy of the Social Sciences, 1*(2), 309–344.

Editorial. (2010). Assessing assessment. *Nature, 465,* 845.

French, B. M., & Koeberl, C. (2010). The convincing identification of terrestrial meteorite impact structures: What works, what doesn't, and why. *Earth-Science Reviews, 98,* 123–170.

Geels, F. W., & Smit, W. A. (2000). Failed technology futures: Pitfalls and lessons from a historical survey. *Futures, 32*(9–10), 867–885.

Gill, J. (2005). An entropy measure of uncertainty in vote choice. *Electoral Studies, 24,* 371–392.

Goodwin, P., & Wright, G. (2010). The limits of forecasting methods in anticipating rare events. *Technological Forecasting and Social Change, 77*(3), 355–368.

Greenberg, J. M. (2004). Creating the "Pillars": Multiple meanings of a hubble image. *Public Understanding of Science, 13,* 83–95.

Gruber, H. E. (1992). The evolving systems approach to creative work. In D. B. Wallace & H. E. Gruber (Eds.), *Creative people at work: Twelve cognitive case studies* (pp. 3–24). Oxford, UK: Oxford University Press.

Guilford, J. P. (1967). *The nature of human intelligence.* New York: McGraw-Hill.

Heuer, R. J. Jr. (1999). *Psychology of intelligence analysis.* Washington, DC: Central Intelligence Agency.

Hsieh, C. (2011). Explicitly searching for useful inventions: Dynamic relatedness and the costs of connecting versus synthesizing. *Scientometrics, 86*(2), 381–404.

Hudson, L. (1966). *Contrary imaginations: A psychological study of the English schoolboy.* New York: Schocken.

Kuhn, T. S. (1977). *The essential tension: Selected studies in scientific tradition and change.* Chicago/London: University of Chicago Press.

Lamb, W. E., Schleich, W. P., Scully, M. O., & Townes, C. H. (1999). Laser physics: Quantum controversy in action. *Reviews of Modern Physics, 71,* S263–S273.

Laudel, G. (2006). The art of getting funded: How scientists adapt to their funding conditions. *Science and Public Policy, 33*(7), 489–504.

MacLean, M., Anderson, J., & Martin, B. R. (1998). Identifying research priorities in public sector funding agencies: Mapping science outputs on to user needs. *Technology Analysis & Strategic Management, 10*(2), 139–155.

Mowery, D., & Rosenberg, N. (1982). The commercial aircraft industry. In R. R. Nelson (Ed.), *Government and technological progress* (pp. 101–161). New York: Pergamon Press.

Newell, A., Shaw, J. C., & Simon, H. A. (1958). Elements of a theory of human problem-solving. *Psychological Review, 65*(3), 151–166.

Perrow, C. (1984). *Normal accidents: Living with high-risk technologies* Princeton, NJ: Princeton University Press.

Pirolli, P. (2007). *Information foraging theory: Adaptive interaction with information.* Oxford, UK: Oxford University Press.

Rothenberg, A. (1987). Einstein, Bohr, and creative-thinking in science. *History of Science, 25*(68), 147–166.

Rothenberg, A. (1996). The Janusian process in scientific creativity. *Creativity Research Journal, 9*(2–3), 207–231.

Rowe, G., Wright, G., & McColl, A. (2005). Judgment change during Delphi-like procedures: The role of majority influence, expertise, and confidence. *Technological Forecasting and Social Change, 72*(4), 377–399.

Scott, G. M., Leritz, L. E., & Mumford, M. D. (2004). The effectiveness of creative training: A quantitative review. *Creativity Research Journal, 16,* 361–388.

Simonton, D. K. (1999). *Origins of genius: Darwinian perspectives on creativity*. New York: Oxford University Press.

Star, S. L. (1989). The structure of Ill-structured solutions: Boundary objects and heterogeneous distributed problem solving. In M. Huhs & L. Gasser (Eds.), *Readings in distributed artificial intelligence 3* (pp. 37–54). Menlo Park, CA: Kaufmann.

Tichy, G. (2004). The over-optimism among experts in assessment and foresight. *Technological Forecasting and Social Change*, *71*(4), 341–363.

Wenger, E. (1998). *Communities of practice—Learning, meaning, and identity*. Cambridge, UK: Cambridge University Press.

Wohlstetter, R. (1962). *Pearl Harbor: Warning and decisions*. Redwood City, CA: Stanford University Press.

Zhao, H., & Jiang, G. (1985). Shifting of world's scientific center and scientists' social ages. *Scientometrics*, *8*(1–2), 59–80.

CHAPTER 3

Subjectivity of Evidence

We have seen in Chapter 2 that it is often hard to change our beliefs or mental models even when critical information is right in front of us. We may overlook such information altogether because it may not be what we are looking for and we probably do not expect to see it. In this chapter, we will investigate the relations between a mental model and potentially critical information that may lead to substantial changes in our mental model.

Any evidence is subjective in nature with respect to our mental models. The value of incoming information can be measured in terms of its strength as evidence that may alter our beliefs. If a particular piece of evidence can truly change the way we see the world, then we would certainly consider its value to be high. In contrast, if a piece of information does not seem to change what we know or what we do, then its value is considered to be low. Since we tend to be biased one way or another, we should be careful when we deem a piece of information as noninformative or worthless.

3.1 THE VALUE OF INFORMATION

The same evidence may be valuable to one person but worthless to another. Different people looking at the same evidence may draw different conclusions from it because it can be viewed and interpreted from different mindsets or mental models.

The discovery of DNA is one example of this kind. In James Watson's autobiography, he described the vital clue he learned from a glimpse of X-ray diffraction images taken by Rosalind Franklin (1920–1958). Franklin's images revealed critical clues of the structure of DNA. Although she suspected that all DNA was helical, she did not want to announce her finding until she had sufficient evidence on the other form of DNA. Franklin's results were shown to Watson, apparently without her knowledge or consent. Watson and Francis Crick were able to pick up the clues

The Fitness of Information: Quantitative Assessments of Critical Evidence, First Edition. Chaomei Chen.
© 2014 John Wiley & Sons, Inc. Published 2014 by John Wiley & Sons, Inc.

FIGURE 3.1   The tip of an iceberg and the shape of the iceberg as a whole.

quickly and advance to their Nobel Prize-winning discovery. To Watson and Crick, Franklin's X-ray diffraction images provided evidence of the helical structure of DNA, which was important enough to keep them on the right track.

Franklin died of cancer at the age of 37, three years before Watson, Crick, and Wilkins won the Nobel Prize for the discovery of DNA in 1962. The significance of her work that led to the discovery of DNA was underestimated. On the other hand, what enabled Watson to immediately realize the importance of Rosalind's X-ray diffraction image and capitalize it for the discovery of the structure of DNA? In many other examples, important clues were overlooked when the right information is literally in front of them. The mind drives what we are looking for.

Figure 3.1 illustrates a profound principle that we will underline with a series of examples and discussions in this book. What is the value of information? The figure illustrates three scenarios of three different shaped icebergs. The only thing that they have in common is the small tip above the water, but the much larger part that differs from others is concealed under the water. If our mental model of an iceberg is constructed on the observation of the tips above the water, then we will probably conclude that they are all the same. If the water level drops further to reveal more information about each of the icebergs, we will notice how different they are. This example suggests that we should consider the value of information in terms of how likely it is to change our current belief.

We review our beliefs when new information becomes available. If the available information is not enough for us to make up our mind, we may gather more information. For example, physicians can order various tests for their patients and make an informed assessment on the situation they are in. Physicians can analyze test results and decide whether additional tests are necessary. In general elections, voters ask questions about candidates' political positions in order to reduce or eliminate uncertainties about choosing candidates.

Thomas Bayes (1702–1761) was an English mathematician and statistician. Today, his name is still being actively mentioned in the literature of many scientific disciplines. According to Google trends, people frequently searched for topics named after him, notably, Bayesian networks, Bayesian analysis, Bayesian statistics, Bayesian inference, and Bayesian learning. Bayesian reasoning is widely used to analyze evidence and synthesize our beliefs. It has been applied to a wide variety of application domains, from interpreting women's mammography for breast cancer risks to differentiating spam from genuine emails.

The search for the USS *Scorpion* nuclear submarine is a popular story of a successful application of Bayesian reasoning. The USS *Scorpion* was lost from the sea in May 1968. An extensive search failed to locate the vessel. The search was particularly challenging because of the lack of knowledge of its location prior to its disappearance. The subsequent search was guided by Bayesian search theory, which is comprised of the following steps:

1. Formulate hypotheses of whereabouts of a lost object.
2. Construct a probability distribution over a grid of areas based on the hypotheses.
3. Construct a probability distribution of finding the lost object at a location if it is indeed there.
4. Combine the two distributions and form a probability distribution and use the new distribution to guide the search.
5. Start the search from the area with the highest probability and move to areas with the next highest probabilities.
6. Revise the probability distribution using the Bayesian theorem as the search goes on.

In the *Scorpion* search, experienced submarine commanders were called in to come up with hypotheses independently of whereabouts of the *Scorpion*. The search started from the grid square of the sea with the highest probability and moved on to squares with the next highest probabilities. The probability distribution over the grid was updated as the search moved along using the Bayesian theorem. The *Scorpion* was found in October more than 10,000 feet under the water, within 200 feet of the location suggested by the Bayesian search.

The Bayesian method enables searchers to estimate the cost of a search at local levels and allows the searchers adapt their search path according to the revised beliefs as the process progresses. This adaptive strategy is profound in a wide range of situations in which we need to integrate new information with our beliefs. If necessary, we may need to change our beliefs altogether.

3.2 CAUSES CÉLÈBRE

Since the early 1990s, lawyers and forensic scientists have turned their attention to computational tools such as Bayesian networks for studying evidence and facilitating inference. Lawyers tend to be more interested in structuring cases as a whole. In contrast, forensic scientists are more interested in evaluating specific scientific evidence.

Bayesian networks have been used in several retrospective studies of complex and historically important cases, that is, causes *célèbre*, such as the Sacco and Vanzetti case, the O.J. Simpson case, the Collins case, and the Omar Raddad case. The use of the Bayesian inference strategy has been a controversial issue. One can easily find diverse opinions on related topics. An interesting observation seems to suggest that the level of interest in computational supports is directly influenced by the complexity and controversy of a case.

3.2.1 The Sacco and Vanzetti Case

Joseph B. Kadane and David A. Schum (1996) provided an extensive probabilistic analysis of evidence in the Sacco and Vanzetti case in their 1996 book published by Wiley. What is striking was Kadane and Schum's following remark: "We live in an age when we are still more adept at gathering, transmitting, storing, and retrieving information than we are at putting this information to use in drawing conclusions from it. To use information as evidence, we must establish its relevance, credibility, and inferential force credentials." What they said 17 years ago is still inspirational today as new fields such as the big data attract much of the attention of the society. "Big data" may be an unfortunately chosen term because it emphasizes the focus on the size rather than the quality of reasoning and the value one can deduce from the data. The real challenge is to uncover insights that would not be detectable with any datasets of less than the "big" size. One example we reviewed earlier in the book is the suspected connection between facial features and chronological illnesses. The connection is so weak that it cannot be detected unless in a large sample. Another example is from astronomy. The data collected by the Sloan Digital Sky Survey (SDSS) made it possible for astronomers to test some of the most fundamental hypotheses for the first time ever. These included a large number of galaxies and quasars, in the range of 200,000 simultaneously, which had substantially advanced our understanding of the universe and led to a profound impact.

The Sacco and Vanzetti case is the most controversial trial in the twentieth century in America, starting with a murder and robbery on April 15, 1920. A paymaster and his guard carrying a factory payroll of US $15,776 were shot and fatally wounded by two men in a small town south of Boston, called South Braintree, Massachusetts. The gunmen fled with a getaway car. A few more people were in the getaway car.

Two Italians, Nicola Sacco and Bartolomeo Vanzetti, were arrested three weeks later by police in a trap that had been set for a suspect in the Braintree crimes. Both were charged for the South Braintree crimes. Vanzetti was also charged with an attempted robbery that took place a few months earlier in the nearby town of Bridgewater.

Vanzetti was tried first on the Bridgewater robbery, which was the lesser of the two charges for him. No one was hurt in the Bridgewater robbery. Vanzetti had a strong alibi by many witnesses. However, testimonies of the witnesses, translated from Italian to English, failed to convince the jury. Sacco and Vanzetti were both anarchists. To avoid revealing his radical anarchist activities, Vanzetti did not take the stand in his own defense, which seriously damaged his case. Vanzetti was found guilty and sentenced for 10–15 years, a much harsher than usual punishment. The harsh punishment was interpreted as politically driven in relation to Vanzetti's anarchist activities rather than the robbery crime he was charged.

After Sacco and Vanzetti were arrested, the police found a draft of a leaflet for an anarchist meeting in Sacco's pocket, with Vanzetti as the main speaker. Both Sacco and Vanzetti were involved in the anarchist movement, and they were believed to be a risk, or at least in support, of revolutionary violence. The initial questioning by the police was focused on their anarchist activities but not specifics of the Braintree robbery. The two lied in response, however, which led to a worse situation. The authorities were convinced that they lied because it was a sign of a "consciousness of guilt" to deny their criminal involvement in the Braintree crime. On the other hand, an alternative explanation was suggested, especially by their supporters, that they lied in an attempt to protect their anarchist friends, given the hysterical atmosphere in the nation toward foreign-born radicals.

Sacco and Vanzetti's new defense lawyer was Fred H. Moore, who was a well-known socialist lawyer. Moore's strategy was to publicize that the prosecution of Sacco and Vanzetti was politically motivated because of their involvement in the Italian anarchist movement. Moore's strategy of publicizing and politicizing the initially little known case made it an international cause célèbre. Some analysts suggested that Moore's strategy was probably backfired. On July 14, 1921, the jury found Sacco and Vanzetti guilty of robbery and murder. Over the next few years, the defense made numerous attempts to appeal to state and federal courts for a new trial, but Judge Webster Thayer firmly rejected all the motions. It was he who sentenced Vanzetti for the Bridgewater case. On April 9, 1927, Sacco and Vanzetti were sentenced to death after all attempts failed.

The two men had shown a degree of dignity that brought them attention of the world. They became symbols of social justice. Their death sentence triggered protests in Tokyo, Paris, London, Sydney, Johannesburg, Mexico City, and Buenos Aires and other cities against the unfairness of their trial. Harvard law professor and future Supreme Court justice Felix Frankfurter argued for their innocence. A three-man committee led by Harvard University's president A. Lawrence Lowell was appointed by Alvan T. Fuller, the governor of Massachusetts to consider whether clemency was warranted in this case. The committee was later known as the Lowell Committee. After interviews with the judge, lawyers, and several witnesses, the Lowell Committee concluded that the trial and judicial process was legitimate "on the whole"—clemency was not warranted.

Sacco and Vanzetti were executed on August 23, 1927, followed by a wave of riots in major cities such as Paris and London. The cause célèbre has raised fundamental questions in America's legal, social, political, and cultural system.

The complexity of the case and a diverse range of plausible and less plausible inter-pretations of evidence spanned an array of opinions across almost all the conceivable combinations. Numerous revisionists subsequently reviewed the case and reached their own and equally diverse verdicts. To name a few, in 1955, Francis Russell was convinced that both were innocent. He revised his verdict in 1962—Sacco was guilty and Vanzetti remained to be innocent. In 1962, New York lawyer James Grossman found Vanzetti innocent. In 1965, David Felix found both of them guilty. In 1982, Francis Russell received a letter from the son of Giovanni Gamberi, who was a member of the group formed in 1920 to arrange the Sacco and Vanzetti's defense. The letter said "Sacco was guilty" and "Vanzetti innocent as far as the actual partic-ipation in the killing."

The cause célèbre provides a classic example for Kadane and Schum to make their case. Kadane and Schum underlined numerous and profound challenges faced not only by the 12 jurors of the Sacco and Vanzetti trial but virtually anyone who demands to know the truth:

> Despite the mass of evidence they saw and heard, the evidence was *incomplete*; in no infer-ential situation do we ever have *all* the evidence. The evidence was also *inconclusive*, hav-ing more than one possible explanation. Some of it was *imprecise* or ambiguous in the sense that its meaning was unclear. In addition the evidence was obtained from sources have *less than perfect credibility*. The witnesses were carefully coached and in many cases gave testimony at trial counter to what they gave in earlier hearings. Finally, the evidence was certainly *dissonant* being in some matters contradictory and in other matters conflicting or divergent. Recognition of these characteristics of evidence is one reason why the stan-dard of proof in criminal trials is: beyond a *reasonable* doubt; and not: beyond *all possi-bility* of doubt. In other words, the inferential task facing jurors is necessarily probabilistic in nature. (Kadane & Schum, 1996, p. 14)

The same piece of evidence is often interpreted in different ways by people with different perspectives or motivations. In the Sacco and Vanzetti case, the prosecu-tion and the defense disagreed on their interpretations of almost every piece of evidence considered. For example, after his arrest, Sacco lied about his recent whereabouts and denied holding radical beliefs. The prosecution suggested that these lies showed a consciousness of guilt. In contrast, the defense argued that he lied because he feared if he told the truth about his radical beliefs, he would be deported. The Lowell Committee later determined that Sacco's lies were consistent with his explanation of being afraid of deportation. After the execution, the defense lawyer Moore revealed that Sacco and Vanzetti had probably gone to Bridgewater to get a car and collect dynamite, but thought that the explanation would be too damaging. Instead, they said they went there to collect anarchist literature, but could not specify the names or addresses of the people he would collect the literature from.

Ballistic evidence was critical in the case, in particular, the ballistic evidence of Bullet 3. For the prosecution, ballistic expert Proctor testified that Bullet 3 was con-sistent with being fired through Sacco's pistol. Expert Van Amburgh noted a scratch on Bullet 3 was "likely" made by a defect of Sacco's pistol. For the defense, two

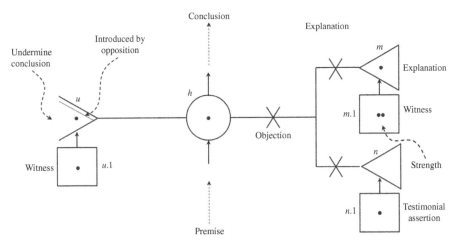

FIGURE 3.2 A sketch of a Wigmore chart.

experts, Burns and Fitzgerald, testified that Bullet 3 could not have been fired from Sacco's pistol. In 1961, a ballistic test conducted at the Massachusetts Police Lab suggested that Sacco's pistol was used to fire Bullet 3.

Kadane and Schum realized the value of a graphical representation to aid the analysis of evidence—Wigmore charts. They adopted Wigmore charts in their analysis of the Sacco and Vanzetti case. Wigmore charts were developed by John Henry Wigmore (1863–1943). He was a U.S. jurist and a pioneer in the law of evidence. He is known as the author of Treatise on the Anglo-American System of Evidence in Trials at Common Law. The purpose of a Wigmore chart is to provide the justification of various uses of evidence by jury or judge at a trial.

A Wigmore chart uses lines, shapes, and text to capture the connection among premises, hypotheses, associated evidence, and counter evidence in terms of witnesses' testimonies and other forms of explanations. Such charts represent paths of reasoning from a premise to a conclusion. Figure 3.2 shows a sketch of a Wigmore chart. A premise h is shown in the middle of the chart. The circle surrounding the h and the arrow pointing up indicate that the premise h challenges a conclusion on the top of the chart. Triangles represent explanations, such as the triangle m in the upper right corner of the chart. Triangles also represent refutations of explanations. A square indicates a testimonial assertion, for example, as the $n.1$ shown in the lower right of the chart, which means that someone testified the assertion under oath. The ultimate belief is denoted within the corresponding shape, for example, "?" for doubt, "•" for belief, "••" for strong belief, or "○" for disbelief. The witness testimony $m.1$ on the right of the chart has a strong belief.

In 2004, Peter Tillers, a professor of law in the Cardozo School of Law at Yeshiva University in New York City discussed the topic of Picturing Factual Inference in Legal Settings in a publication. He focused on the role of visual representations of entities and relationships that might be involved in an inference process in a legal context, especially in trials. A major challenge he identified was the tension between the global and granular perspectives of our reasoning and inference needs. On the one hand, we need to access to the big picture of the entire situation. On the other hand, we need to pay attention to details at a particular stage of the inference. He emphasized the critical role of tacit and implicit hypotheses and evidence in the overall inference process because we can only pay attention to a small number of issues simultaneously. We have discussed some of the same issues in Chapter 2, for example, in the example of the wrong patient case.

He criticized Wigmore's chart methods and argued that Wigmore's charts are not widely adopted because they tend to be too complicated. In terms of the potential benefits of accessing the big picture, Tillers argued that one way to increase people's ability to retrieve and remember information is to organize information in a way that is meaningful to the user. He distinguished synthetic and tacit processes but stresses both are necessary in our inference process. Although he just discussed the issues concerning why it would be nice to have visualizations that could facilitate short term, his idea was still in the process of conceptualization. There was no empirical test.

Kadane and Schum in their analysis of the Sacco and Vanzetti case combined the structural insights of Wigmore and a probabilistic inference approach based on Bayesian inference. They referred to their integrated approach as a conceptual microscope to examine the inference strength of a complex body of evidence.

To Kadane and Schum, the value of Wigmore's structural methods is in its ability to uncover sources of uncertainty or doubt in constructing complex networks of arguments from a mass of evidence. The clarification of sources of doubt is critical for jurors or anyone in a comparable situation to come to their own conclusions whether it would be beyond a reasonable doubt. In parallel, Bayes's rule enables one to trace the update of prior beliefs in light of new evidence and provides a way to communicate the reasoning process among people who may have multiple conceptualizations of the situation. By marshaling evidence with the probabilistic inference lens, they developed a conceptual and analytic microscope.

Figure 3.3 shows the major argument of the prosecution, the probandum,[1] in the middle of the chart. The probandum states that Sacco and Vanzetti were guilty of first-degree murder in the Braintree robbery. To establish the validity of the statement, one would need to be convinced through propositions 1–3. Proposition 1 stated that someone died of gunshot wounds. Proposition 2 stated that the victim was in possession of a factory payroll when he was shot. Proposition 3, the most challenging one, stated that it was Sacco who intentionally fired at the victim. One can expand each proposition with a structure of a finer granularity such as the one shown in Figure 3.2. Wigmore's methods were revitalized by Terrence Anderson and William Twining in their 1991 book entitled *Analysis of Evidence: How to Do Things with Facts Based upon Wigmore's Science of Judicial Proof.*

---

[1] Probandum: Latin for "that which is to be proved."

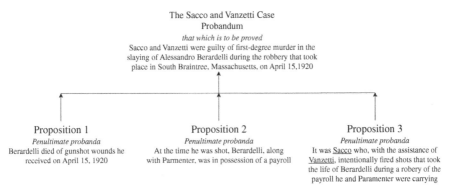

FIGURE 3.3   Prosecution's top-level propositions in the Sacco and Vanzetti case.

3.2.2   The O.J. Simpson Case

The O.J. Simpson case is another widely publicized and controversial trial in the United States. Former football star O.J. Simpson's ex-wife, Nicole Brown Simpson, and her friend Ronald Lyle Goldman were found murdered on June 13, 1994. On June 17, 1994, an eventful 50-mile low-speed chase of O.J. Simpson was covered live on television by ABC news. Thousands of people were watching the chase unfold on overpasses along the way. Domino's Pizza later reported "record sales" of pizza delivery during the televised chase. O.J. Simpson surrendered to the police after the chase ended. He pleaded not guilty to both murders.

The controversial and extensively publicized criminal trial of the murder case lasted 252 days. Despite a "mountain of evidence" presented by the prosecution, Simpson's defense lawyers convinced jurors that Simpson's charge was not beyond reasonable doubt. The announcement of the verdict of the trial attracted the attention of many people. Again, Domino's Pizza noticed a spike in their sales 15 minutes before the afternoon verdict reading of the trial on October 3, 1995. The surge ended abruptly when the verdict was announced. For the next five minutes, not a single pizza was ordered in the entire country.[2] The jury's verdict was not guilty.

Both the Brown and Goldman families sued Simpsons subsequently for damages in a civil trial. The jury of the civil trial found Simpson liable for damages in the wrongful death of Goldman and battery of Brown.

Paul Thagard, a Canadian philosopher at the University of Waterloo, is most well known for his work on the theory of explanatory coherence and using it to explain conceptual revolutions in science. According to his theory, the acceptance or rejection of a new idea boils down to whether the new idea has enough explanatory coherence power. In an article published in 2003, he applied his theory to the analysis of the O.J. Simpson case in an attempt to address the question why the jury found him not guilty.

He compared four competing explanations for why the jury of the O.J. Simpson murder trial decided that he was not guilty. The four alternative explanations were

[2] http://articles.chicagotribune.com/1995-12-28/business/9512280130_1_pizza-meter-domino-s-pizza-underwear

proposed through four different methods, namely, the explanatory coherence theory, Bayesian probability theory, wishful thinking, and emotional coherence. The question was to find which of the four methods would lead to the same verdict as the jury did. Whichever method could reconstruct the actual verdict of the jury would be considered the most plausible answer as to why the jury found him not guilty. The jury of a criminal trial is advised to make their decision on the ground whether it is beyond the reasonable doubt that the defendant committed the crime.

In terms of the *explanatory coherence* explanation, the jury deemed Simpson not guilty because it did not seem plausible that he committed the murder. The plausibility is determined by explanatory coherence.

The second explanation was constructed on the basis of *Bayesian probability theory*. The jury found O.J. Simpson not guilty because the probabilities calculated by Bayes's theorem were not strong enough to convince the jury.

The third explanation was based on a *wishful thinking* hypothesis that the verdict was what the jury wanted all along regardless of the arguments. This explanation is probably not as convincing as the other alternative explanations. Besides, even if the wishful thinking explanation is indeed the case, we still won't have much to learn.

The fourth explanation is motivated by *emotional coherence*. It focused on an interaction between emotional bias and explanatory coherence.

Each of the four explanations corresponds to a process of reasoning and inference that can be simulated computationally, except the one for the wishful thinking explanation. Thagard studied the results of computational simulations to find out if they produced the same verdict as the jury. As it turned out, emotional coherence duplicated the verdict. Thus, the reasoning process behind emotional coherence was considered the most plausible explanation for why the jury declared Simpson not guilty.

From the perspective of social psychology, our cognition has two different kinds, depending on whether motivations and emotions are involved or not. Cognition that involves emotions is hot, whereas cognition that does not involve emotions is cold. Among the four explanation processes, the explanatory coherence theory and Bayesian probabilistic inference are cold cognitive because they do not take into account rationales or emotions. In contrast, wishful thinking is in essence emotional and it is therefore hot cognition. Finally, the emotional explanatory coherence explanation is a mixture of hot and cold cognition. Perhaps Thagard should classify it as warm cognition.

In the O.J. Simpson trial, the prosecution collected a "mountain of evidence" to build the case. For example, a glove with blood on it, found in Simpson's backyard, matched the glove found at the crime scene. He was abusive and jealous. He seemed to have a motive. On the other hand, the defense did a great job to undermine the arguments made by the prosecution. Simpson was asked to put the glove on in front of everyone, but it didn't fit. There was a cut on Simpson's hand, but nothing on the glove matched the cut. Which side's story was more compelling to the jury? Which fine detail was critical to the jury that the case was beyond the reasonable doubt?

According to Thagard's theory, what it takes is a coherent story that can explain various evidence and possible connections. Many people believed that Simpson was guilty, but the jury apparently wasn't convinced by the case presented by the prosecution.

Thagard had developed a method of explaining what determines the acceptance of competing scientific theories. He applied the theory, explanatory coherence, to the Simpson case. The explanatory coherence theory suggests that an argument is likely to be accepted if it provides the best possible coherence across all the evidence and interpretations. In other words, the best candidate is expected to have the highest level of coherence than alternative arguments.

The level of coherence is measured as follows. Given two propositions p and q, their explanatory coherence is a relation between them, $coh(p, q)$. The relation is symmetric, so $coh(p, q) = coh(q, p)$. If propositions p and q are contradictory to each other, then they have the lowest level of coherence, or equivalently, they are not coherent with each other. If sim is a similarity function, then $coh(p, q) = coh(sim(p), sim(q))$, which means that the similarity function preserves the coherence level. Explanations P and Q are explanatorily connected if one explains the other or if they explain something else together.

Given a system that consists of multiple propositions, the acceptance of a new proposition is determined by how strongly it is coherent with the existing propositions. In terms of mental models and the OPERA's over-the-speed-of-light finding we have discussed in Chapters 1 and 2, traveling over the speed of light contradicts Einstein's relativity theory. It would not be accepted due to its incoherence. On the other hand, the later announcement that the measurement was flawed is consistent with other propositions that one can derive from the relativity theory. It has been accepted uneventfully. Since it didn't add any new propositions or connections, it didn't cause any changes to the belief system after all.

Thagard implemented a computational model called ECHO to test the explanatory coherence. Hypotheses and evidence are represented as interconnected units. Connections between units are characterized by two types of links, either excitatory or inhibitory. An excitatory link connects coherent units, whereas an inhibitory link connects incoherent units. Units connected in this way form a network. A unit can be accepted or rejected. An acceptance of a unit is represented by a positive activation of the unit, whereas a rejection is represented by a negative activation of the unit. The activation of a unit is controlled by a threshold for firing and a decay factor. A spreading activation process is defined on these graphs such that all the units are activated based on how they are connected with neighboring units and the type of connections between them. The structure of the network is updated by the spreading activation process.

The spreading activation technique is commonly used to find the shortest path between two artificial neurons in a neural network. The process starts by activating both neurons simultaneously, which will trigger the activations of neurons in the network along the structure of the network. The spreading activation process functions like sending out multiple agents to search in different directions simultaneously. The shortest path is identified when two agents originated from different starting points meet on a particular neuron.

Thagard's theory is built on a network of propositions and evidence of an underlying phenomenon. Different models of the same phenomenon can be represented by differently structured networks. Each model is evaluated in terms of the status of its

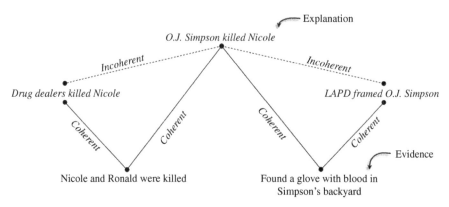

FIGURE 3.4   Coherent and incoherent explanations and evidence.

member units. Each node in a network of this kind may have one of the two values: accepted or rejected. Each link in the network has a positive weight. The number of coherent links in the network represents the overall explanatory coherence of a model. According to Thagard, a model with the highest number of coherent links is the best candidate.

In the Simpson case, the goal of his defense team was to undermine the prosecution's arguments by convincing the jury that there are other plausible interpretations of the evidence, and therefore, the case is not beyond a reasonable doubt. We illustrate how this strategy works in Figure 3.4, in which propositions are connected with evidence through coherent or incoherent links.

The prosecution wanted to convince the jury beyond a reasonable doubt that "O.J. Simpson killed Nicole." The defense lawyers, on the other hand, wanted to convince the jury that there could be other explanations of the evidence presented by the prosecution. For example, an alternative hypothesis is that drug dealers killed Nicole. Another alternative interpretation is that someone else killed Nicole, but the LAPD framed O.J. Simpson instead. If the defense could demonstrate that other explanations have a reasonable chance to be true, then it would be less likely for the jury to accept the prosecution's argument.

The proposition that O.J. Simpson killed Nicole is consistent with the evidence that a glove with blood was found in Simpson's backyard. Similarly, the proposition that the LAPD framed him is also consistent with the evidence. The consistency, that is, the coherence, is shown in solid lines. In contrast, incoherent explanations are connected with dashed lines. If the proposition that O.J. Simpson killed Nicole is true, then it would activate more evidence than other propositions in this simplified example. According to the theory of explanation coherence, that would be the most compelling line of argument to the jury.

When Simpson tried to put on the glove in front of the jury, it didn't fit. This apparently contradicted to the proposition that the prosecution tried to establish beyond a reasonable doubt. It also substantially increased the creditability of the proposition that the LAPD framed Simpson (Figure 3.5). Although the prosecution

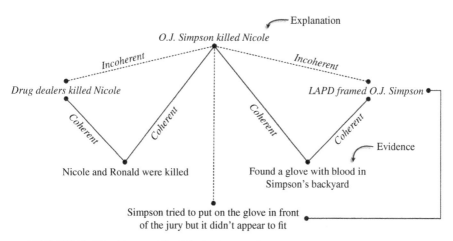

FIGURE 3.5 "O.J. Simpson killed Nicole" has more incoherent links than other propositions.

came up with their own explanations why it didn't fit, for example, Simpson did not take his medicine for arthritis the day before, the damage was done.

Interestingly, Thagard's computational model suggested that the most plausible explanation was that both propositions are true. In other words, O.J. Simpson killed Nicole and the LAPD framed him.

Unlike the jury, Thagard's ECHO system found Simpson guilty. What does the different verdict mean? How does it address the original question on why the jury found Simpson not guilty? Thagard concluded that the cold cognition model apparently did not track the thinking of the jury. What was missing in the explanatory coherence theory?

Next, he considered another cold cognition approach—Bayesian probabilistic inference. From a probabilistic perspective, the jury's question was, "what is the probability that Simpson was the murderer given all the evidence?"

There is a fundamental difference between the probability of the murderer and the probability that a coin will land with its head side up. While you can toss a coin endlessly and make observations forever, each murder case is unique and it cannot be replayed afterward. In the former context, the concept of probability is objective. In contrast, in the latter context, the notion of probability is subjective. It quantifies the plausibility of an event. Scholars distinguish these two types of probabilities. The latter one is often called Bayesian probability.

Bayesian rules represent a fundamental realization: $P(H|E) \times P(E) = P(E|H) \times P(H)$, where H represents a hypothesis and E represents evidence. In Simpson's criminal trial, the question was to estimate the value of $P(H|E)$, that is, given the evidence, what is the probability that Simpson was the murderer? In other words, $P(H|E) = P(\text{guilty}|\text{evidence})$.

According to Bayes's rules $P(\text{guilty}|\text{evidence}) = P(\text{evidence}|\text{guilty}) \times (P(\text{guilty})/P(\text{evidence}))$. To estimate the probability that would lead to a verdict, $P(\text{guilty}|\text{evidence})$, it is necessary to figure out three other probabilities. $P(\text{guilty})$ is the probability before we are aware of evidence, which is the prior probability. It represents our initial belief. P(guilty|evidence) is the probability of an event given new evidence. It is the posterior probability, which represents how our previous belief is affected in light of new evidence. P(evidence|guilty) is the likelihood of the evidence if Simpson is indeed guilty. P(evidence) is the probability of observing the evidence independently of anything else. This reasoning is a probabilistic inference process.

The foundation of probabilistic reasoning has been questioned since it was introduced. The most worrying criticism challenges the assumption that the beliefs of human beings are governed by the laws of probability. In fact, there are numerous examples of how human beings performed poorly in Bayesian-style thinking. For example, even experienced physicians performed incompetently in estimating the probability that a patient has cancer based on a positive lab test.

Thagard's criticism on the applicability of Bayesian networks in criminal trials is based on the observation that members of the jury would have no idea how to figure out probabilities such as P(found a glove with blood|O.J. Simpson killed his ex-wife), the probability of finding a glove with blood given O.J. Simpson killed his ex-wife. Thus, the members of the jury would have no way to determine whether an estimated posterior probability is reasonable or not.

Another weakness of the Bayesian approach is that it may give the false impression that a probability level is more precise than the expression of "beyond a reasonable doubt." Indeed, it is widely believed that the beyond-a-reasonable-doubt guidance is given intentionally in its current form rather than a quantitative value.

To find out what verdict would be generated by Bayesian probabilities, Thagard used JavaBayes, a Bayesian simulator, to work with the probabilities of various propositions and evidence in the Simpson trial. In JavaBayes, the user needs to come up with some of the probabilities as the initial input. Thagard found it very hard for him to come up with as many as 60 conditional probabilities. Furthermore, he found it even harder to justify the probability values he gave. For example, if your estimate of the value of P(Simpson killed Nicole|Simpson was abusive) is 0.2, how would you justify that it should not be 0.1 or 0.3? In other words, there seemed to be a logical or psychological gap between a proposition stated in a natural language and a quantitative estimate of its probability.

Despite the challenges, JavaBayes eventually produced some interesting results: P(Simpson was murderer|evidence)=0.72, P(drug killings|evidence)=0.29, and P(LAPD framed Simpson|evidence)=0.99. The results of Bayesian inference are consistent with that of the explanatory coherence model. Simpson was very likely to be guilty, and it was almost certain that LAPD had framed him, provided the 60 conditional probabilities entered earlier on were reliable.

Thagard's experience showed that the explanatory coherence process is much easier to operate than the Bayesian inference one, although none of the two cold models replicated the jury's verdict. The two models still didn't quite catch the reasoning process of the jury.

To identify what else may have influenced the jury's reasoning process, Thagard turned his attention to the role of valence in activating a proposition in the network of hypotheses and evidence. He noticed that many of the black jurors had a positive sentiment toward Simpson, but their emotion toward the LAPD was negative. If a juror had a positive attitude toward Simpson, it would be harder for him or her to accept the proposition that Simpson killed his ex-wife. In addition, if a juror had a negative sentiment toward the LAPD, the proposition that they framed Simpson would seem to be a more plausible explanation overall. Since such emotional biases may influence an individual's judgment on a proposition, he called them the valance of a proposition. To test the effect of taking the valance of a proposition into account, Thagard introduced two more emotionally oriented propositions: Simpson is good, and the LAPD is good. He then assigned a positive valence of 0.05 to the proposition that Simpson is good and a negative valence of −0.05 to the proposition that the LAPD is good. The revised model rejected the hypothesis that O.J. Simpson killed Nicole. The finding was consistent with the conclusion that the jury had reached.

Thagard argued that the ability of the revised model to reach the same verdict as the jury did indicates that the model revealed insights into the role of emotional factors in the jury's reasoning process. It provides a plausible scenario that the jurors might have gone through a similar path of thinking, which is characterized, at least on the surface, by a mixture of logic, probability, and emotion. In later chapters, we will encounter research topics on the relationship between underlying combinations at the genotype level and visible features at the phenotype level. Similar combinations of genes at the genotypic level may lead to substantially different phenotypes. These are the properties of a complex system.

One of the lessons from Thagard's study is that people who seem to have strong emotional bias should not be members of a jury. It might sound ironic that emotional bias appears to play a critical role in what is supposed to be a rigorous reasoning process. On the other hand, humans all have emotions. Some may hide them deeply, while others may express them. As Kadane and Schum insightfully summarized, the complexity of the reasoning process is underlined by incomplete information, uncertain observations, ambiguous descriptions, inconsistent testimonies, carefully coached witnesses, and many other sources of uncertainty. There are all sorts of gaps in multiple dimensions, from logical and psychological to social and emotional.

In a nutshell, Thagard's study demonstrates that handling Bayesian inference may be beyond the reach of ordinary members of a jury. The theory of explanatory coherence may reveal insights that individuals may not realize; namely, the claims that Simpson killed Nicole and that he was framed by the LAPD could be simultaneously true. In addition, the theory can be expanded to incorporate the impact of emotional bias into the system in terms of the valence of a proposition. Nevertheless, the question remains whether or not the members of the jury had gone through the same process of reasoning even though the outcomes were consistent.

Obviously, the emotionally enhanced explanatory coherence theory offers the best explanation of the jury's verdict among the four options, but is it still possible that there are other explanations that we just don't realize it? In terms of the feasibility of the Bayesian inference approach, it does sound like a daunting task to come up with 60 conditional probabilities upfront. More importantly, the justification of such decisions is not easy to communicate and share with others. Thus, the overreliance on these initial estimates may well become the Achilles's heel of this approach.

What would Darwin say if he were asked to deal with this problem? Perhaps an evolutionary strategy is worth considering by introducing multiple iterations of learning and reducing the reliance on premises that cannot be empirically justified. Each generation may improve on the structure of earlier generations through the process of evolution. An initially large gap would not do much harm on the final decisions because evolution may further reduce the gap and select the most salient and the fittest combinations. Indeed, such strategies would be consistent with the principle of Bayesian search we discussed earlier in this chapter. It seems to be more reasonable to allow our beliefs updated as new information comes in than insist on betting on a hit or miss of a single attempt.

3.2.3   Ward Edwards's Defense of Bayesian Thinking

Criticisms of Bayesian thinking did not begin with Thagard's study. In 1991, Ward Edwards wrote a 50-page paper on the use of Bayesian inference in legal cases. Major criticisms of the Bayesian approach appeared as early as the 1970s, notably by Michael Finkelstein, William Fairley, and Laurence Tribe. Jonathan Koehler and Richard Lempert made counterattacks. Edwards began with the Collins case before he addressed the criticisms of using Bayesian reasoning in formal legal settings.

The Collins case involves a second-degree robbery. Janet and Malcolm Collins were convicted by a Los Angeles jury in the case. Malcolm appealed his case to the Supreme Court of California on the ground that the prosecution used the theory of probability during the trial and that was damaging to the rights of the defendant.

The robbery in the Collins case took place on July 18, 1964, in the City of Los Angeles. The victim, Mrs. Juanita Brooks, was walking along an alley. As she was picking up an empty carton from the ground, someone suddenly pushed her down to the ground. Then she saw a young woman running from the scene. Mrs. Brooks found her purse missing, which had about US \$40 in it. A witness testified that he saw a woman with a dark blond ponytail run out of the alley and enter a yellow car parked across the street from him. He noticed that the driver was a black male with a mustache and beard.

The prosecutor in the Collins case used a set of probabilities as part of a Bayesian inference approach. For example, the probability of a girl with blond hair was estimated to be 0.333. The probability of a man with mustache was estimated to be 0.25. The probability of a couple in the same car from different races was believed to be 0.001. The prosecutor demonstrated to the jury how they calculated the probability that the robbery was committed by someone else rather than the Collins. The prosecutor multiplied these probabilities and concluded the probability is about one out of 12 million. In other words, it would appear to be almost certain that the Collins committed the robbery. It would be equivalent to say that this is well beyond any reasonable doubt.

The probabilities and how they were used by the prosecution sparked many criticisms. A key criticism is that the prosecutor did not provide evidence to justify these probability estimates. Thagard noted the same problem in his study when he had to come up with prior probabilities on numerous occasions and found it hard to justify what he got. Edwards defended the use of Bayesian networks as practical tools for thinking about evidence and proof. He argued that the balance between the gains and pains from using Bayesian tools had been dramatically changed by the advances of new technologies. He believed that practicing lawyers should be asking how they can exploit these tools in or out of court. More specifically, Edwards explained that probabilities are judgment-based compressions of information. Given a hypothesis H, the probability of $P(H)$ compresses what we currently know about how likely H is to occur or have occurred. $P(H)$ is therefore both prior and posterior. It is prior to whatever new information we can think of how likely H is. It is posterior to the relevant information on which it is based. Edwards claimed, "All probabilities are both posterior and prior in exactly this sense." He suggested that the odds-likelihood ratio version of Bayes's rules is helpful.

For hypotheses H and $\sim H$, the posterior odds for H, given evidence E is $\Omega(H|E) = P(H|E)/(P|\sim E)$. $\sim E$ is the negation of E, just as $\sim H$ is the negation of H. The prior odds, which are the odds before evidence E was observed, is $P(H)/P(\sim H)$. The likelihood ratio $L(E|H) = P(E|H)/P(E|\sim H)$ is the ratio of the probability that E would be observed if H were true to the probability that E would be observed if H were not true. It follows:

$$\Omega(H|E) = L(E|H) \times \Omega(H)$$

Likelihood ratios associated with evidence E remind us that the function of an observation is to help discriminate competing hypotheses. The likelihood ratio, more commonly its logarithm, specifies numerically how helpful the observation is. The significance of an observation is therefore determined by its potential to change our opinion. Edwards stressed the importance of performing sensitivity analysis to find out how a model responds to change in its structure or individual inputs, or both. We will discuss the topics of sensitivity analysis and evolution thinking further later in the book.

3.3   THE DA VINCI CODE

Sometimes it is not sufficient for us to know about people's opinions because it still may not be clear to us what we can do about them and what actions would be appropriate. For example, a customer bought the bestseller from Amazon, but he didn't like it and gave a one-sentence review on Amazon. Now we know that he didn't like it and there is little we can do about it. We probably don't even have enough information to start a conversation. It would be much more informative if we can find out why he didn't like the bestseller. Furthermore, given that it is a bestseller, there must be a lot of people like it. What could make such differences in their opinions?

In the O.J. Simpson example earlier in the chapter, the jury's verdict did not seem to be obvious in light of the "mountains of evidence" against Simpson. There was a seemingly substantial gap between the mountains of evidence and the jury's not guilty verdict. Because of the level of complexity involved, any attempt to make a single jump over the huge gap in cognition is unlikely to succeed. How did Thagard bridge the gap? He introduced a number of touchstones so that the reader can take a series of smaller steps instead. The reader finds each of the small steps convincing and justifiable. In the following example, we demonstrate how we may break down a like or dislike opinion into finer-grained opinions so that we can better understand specific reasons for an opinion.

3.3.1 Positive and Negative Reviews

We will now analyze positive and negative reviews of a controversial bestseller book *The Da Vinci Code*. Customer reviews of *The Da Vinci Code* were retrieved from Amazon.com. Amazon customer reviews are based on a five-star rating system, where a five-star rating is the best and a one star is the worst. We recoded reviews with four or five stars as positive reviews. Reviews with one or two stars are recoded as negative. Reviews with three stars are not considered in this analysis. The positive-to-negative review ratio is about 2:1 (Figure 3.6).

Understanding conflicting book reviews has implications far beyond books, ranging from opinions on merchandise, electronic devices, information services, to opinions on wars, religious, and environmental issues. Advances in this area have the potential of making substantial contributions to the assessment of the underlying credibility of evidence, the strength of arguments, diverse perspectives, and expectations.

The Da Vinci Code is controversial. It attracted many positive reviews and negative reviews. What made it a bestseller? Which aspects of the book were favorably

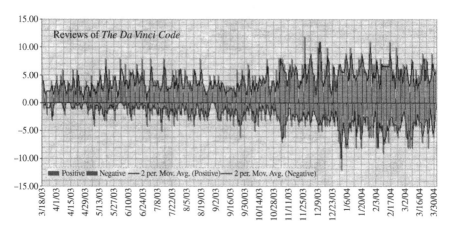

FIGURE 3.6 The distribution of customer reviews of *The Da Vinci Code* on Amazon.com within the first year of its publication (March 18, 2003–March 30, 2004). Although positive reviews consistently outnumbered negative ones, arguments and reasons behind these reviews are not apparent. Source: From Chen, Ibekwe-SanJuan, SanJuan, & Weaver, 2006.

reviewed? Which aspects were criticized? More generally, will we be able to apply the same technique to other bestsellers, movies, cars, electronic devices, innovations, and scientific work? Ultimately, what are the reasons and turning points behind a success, a failure, a controversial issue, or conflicting information from multiple perspectives?

Sentiment analysis aims to identify underlying viewpoints based on sentimental expressions in texts. Pang and Lee (2004) presented a good example of classifying movie reviews based on sentiment expressions. They used text-categorization techniques to identify sentimental orientations in a movie review and formulated the problem as finding minimum cuts in graphs. In contrast to previous document-level polarity classification, their approach focuses on context and sentence-level subjectivity detection. The central idea is to determine whether two sentences are coherent in terms of subjectivity. It is also possible to locate key sentimental sentences in movie reviews based on strongly indicative adjectives, such as *outstanding* for a positive review or *terrible* for a negative review. Such heuristics, however, should be used with considerable caution because there is a danger of overemphasizing the surface value of such cues out of context.

The majority of relevant research is built on the assumption that desirable patterns are prominent. Although this is a reasonable assumption for patterns associated with mainstream themes, there are situations in which such assumptions are not viable; for example, detecting rare and even one-time events and differentiating opinions based on their merits rather than the volume of voice.

During the time frame of the analysis, there were about twice as many positive reviews as there were negative reviews. The average length of positive reviews was approximately 150 words and 9 sentences, whereas negative reviews were slightly longer, 200 words and 11 sentences on average. These reviews are generally comparable to news and abstracts of scientific papers in terms of their length (Table 3.1).

Our goal is to verify the feasibility of predicting the positions of reviews with a small set of selected terms. In addition, we expect decision trees to serve as an intuitive visual representation for analysts to explore and understand the role of selected terms as specific evidence in differentiating conflicting opinions. If we use selected terms to construct a decision tree and use the positive and negative categories of reviews as leaf nodes, the most influential terms would appear toward the root of the tree. We would be able to explore various alternative paths to reach positive or negative reviews.

Table 3.1 Statistics of the reviews

Corpus	Reviews	No. of chars (mean)	No. of words (mean)	No. of sentences (mean)
Positive	2092	1,500,707 (717.36)	322,616 (154.21)	19,740 (9.44)
Negative	1076	1,042,696 (969.05)	221,910 (206.24)	12,767 (11.87)
Total	3168	2,543,403	544,526	32,507

In order to put the predictive power of our decision trees in context, we generate additional predictive models of the same data with other widely used classifiers, namely, the naïve Bayesian classifier and support vector machine (SVM) classifier. We expect that although decision trees may not give us the highest prediction accuracy, it should be a worthwhile trade-off given the interpretability gain.

Reviews are first processed by part-of-speech tagging. We include adjective as part of the phrases to capture emotional and sentimental expressions. Log-likelihood tests are then used to select terms that are not purely high frequent, but influential in differentiating reviews from different categories. Selected terms represent an aggressive dimensionality reduction, ranging from 94.5% to 99.5%. Selected terms are used for decision tree learning and classification tests with other classifiers.

SVM can be used to visualize reviews of different categories. Each review is represented as a point in a high-dimensional space S, which contains three independent subspaces S_p, S_q, and S_c: $S = S_p \oplus S_q \oplus S_c$. S_p represents a review purely by positive reviews. Similarly, S_q represents a review in negative review terms only, and S_c represents a review common in both categories. In other words, a review is decomposed into three components to reflect the presence of positive review terms, negative review terms, and terms that are common in both categories. Note that if a review does not contain any of these selected terms, then it will not have a meaningful presence in this space. All such reviews are mapped to the origin of the high-dimensional space, and they are excluded from subsequent analysis.

The optimal configuration of the SVM classifier is determined by a number of parameters, which are in turn determined based on a k-fold cross-validation (Chang & Lin, 2001). This process is known as model selection. A simple grid search heuristic is used to find the optimal parameters in terms of the average accuracy as to avoid the potential overfitting problem.

3.3.2   Decision Trees

Two decision trees are generated to facilitate tasks for differentiating conflicting opinions. The top of the tree contains terms that strongly predict the category of a review, whereas terms located in lower part of the tree are relatively weaker predictors.

In the 2003 decision tree, the presence of the term *great read* predicts a positive review (Figure 3.7). Interestingly, if a review does not mention *great read* or *Robert Langdon*, but instead talks about *Mary Magdalene*, it is more likely to be a negative review. Similarly, the branch at the lower right corner shows that if a review mentions both *Mary Magdalene* and *Holy Grail*, then it is also likely to be a negative review. In comparison, in the 2004 decision tree, the term *first page* predicts a positive review (Figure 3.8).

The accuracy of this decision tree relies on the 30 most active terms, that is, they have the greatest number of variants in reviews. For example, *robert langdon story* has 250 variants in reviews, 85% of which are positive. Similarly, terms such as *opus dei website, millennium-old secret society*, and *historical fact revelation* have more than 100 variants in reviews of which 66% are positive.

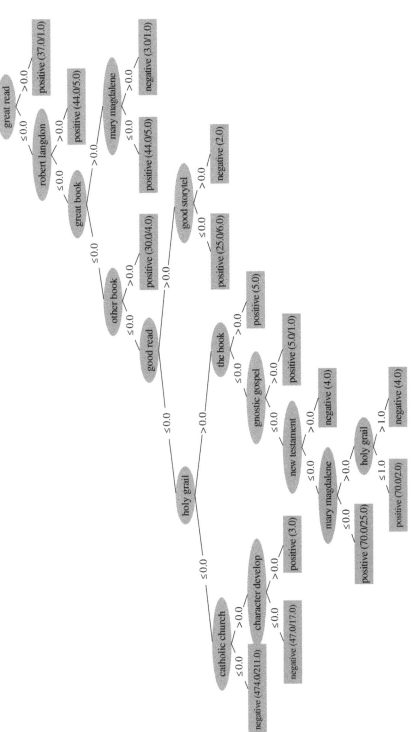

FIGURE 3.7 A decision tree representation of terms that are likely to differentiate positive reviews from negative reviews made in 2003. Source: From Chen et al. (2006).

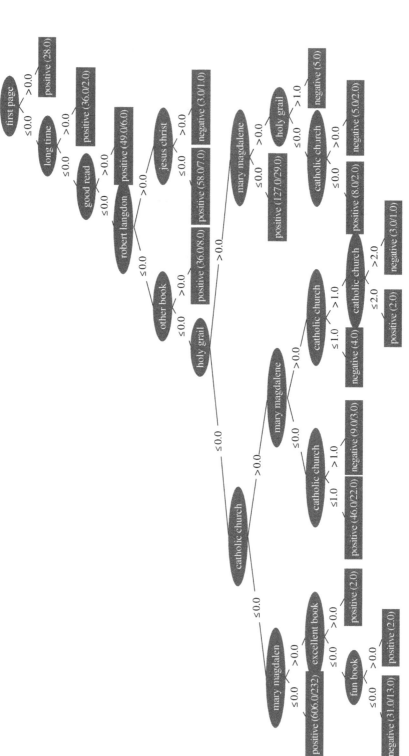

FIGURE 3.8 A decision tree based on reviews made in 2004. Source: From Chen et al. (2006).

Browsing terms not included in the decision tree model is also informative. For example, each of the terms *anti-christian*, *secret grail society blah blah blah*, and *catholic conspiracy* has only six variants in reviews, all negative, identifying readers shocked by the book.

As it turns out, each of the terms like *jesus christ wife*, *mary magdalene gospel*, *conspiracy theory*, and *christian history* have more than 50 variants that are almost evenly distributed between positive and negative reviews.

The perspective of term variation helps to identify the major themes of positive and negative reviews. For negative reviews, the heavy religious controversies raised by the book are signified by a set of persistent and variation rich terms such as *mary magdalene*, *opus dei*, and *the holy grail*, and none of these terms ever reached the same status in positive reviews. Much of the enthusiasm in positive reviews can be explained by the perspective that the book is a work of fiction rather than scholarly work with discriminating terms such as *vacation read, beach read*, and *summer read.*

The same technique is applicable to analyze customer reviews on other types of products, such as the iPod (Figure 3.9), and identify emerging topics in a field of study (Figure 3.10). Figure 3.10 depicts a decision tree representation of noun phrases that would identify a new topic or an old topic in terrorism research. As shown in the decision tree, the term *terrorist attack* is an old topic with respect to the time frame of 2004–2005. In contrast, *mental health* is a new topic. The *risk assessment* of *biological weapon* in particular is a new topic (at the time of data analysis).

In summary, the analysis of the reviews of *The Da Vinci Code* illustrates the nature of conflicting opinions. One of the primary reasons people held different opinions is because they view the same phenomenon through different perspectives. The examples regarding the iPod Video and terrorism research illustrate the potential of the approach for a wider range of applications.

3.4 SUPREME COURT OPINIONS

In criminal trials, both the prosecution and the defense put forward their own arguments supported with carefully selected evidence to convince the jury that their argument is more plausible than their opponent. The details of the decisions made by courts and justifications of judges provide a rich source of information that demonstrates specifically how evidence is interpreted and why a particular line of argument is valid.

The Supreme Court of the United States is the highest federal court in the United States. It has ultimate appellate jurisdiction over all federal courts and state courts. It is the final interpreter of federal constitutional law. The Supreme Court consists of a chief justice and eight associate justices. Each justice has one vote. Sometimes, the justices are in agreement and decide on cases with unanimous votes, whereas other times cases are decided by 5–4 votes.

Decisions made by the Supreme Court on cases are documented in U.S. Supreme Court Opinions, in which justifications are given by the justices. When the court

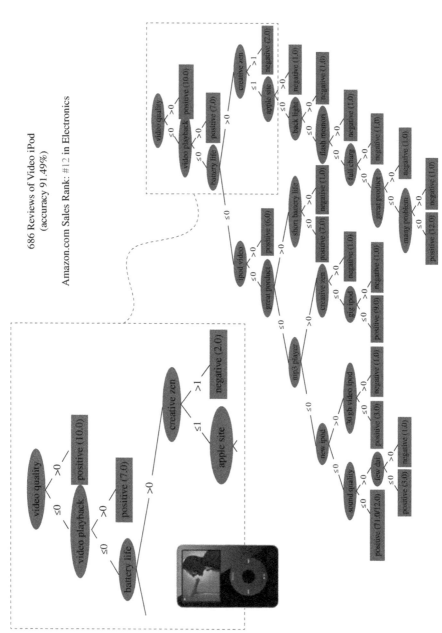

686 Reviews of Video iPod
(accuracy 91.49%)

Amazon.com Sales Rank: #12 in Electronics

FIGURE 3.9 An opinion differentiation tree of Video iPod reviews.

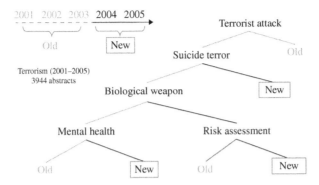

FIGURE 3.10   Representing emerging topics in 3944 abstracts of publications on terrorism.

Table 3.2   **Statistics of the set of U.S. Supreme Court Opinions**

Attribute	Value
Cases	61,509
Courts	31
Year	238
Range	1754–2005
Citations	796,980
Unique citing cases	25,616
Unique cited cases	248,692

considers a case, a significant source of evidence is how relevant cases were ruled in the past. References to decisions on previous cases are known as case citations. Case citations are documented in series of books called reporters or law reports. Similar practices of citation are found in scientific publications and patent applications, although they serve distinct purposes.

We use a publicly available set[3] of the U.S. Supreme Court Opinions to illustrate what one may learn from these types of analyses of evidence. Table 3.2 summarizes the dataset, including 61,509 cases, 31 courts, and 796,980 case citations. The dataset is not complete. We use it for demonstration purposes.

Table 3.3 shows an example of top 100 most cited cases within the dataset. The leftmost column is the total number of times the case is cited. The second column from the left is the reference to the case. The third column is the name of the case, also called parties. The most cited case here is *United States v. Detroit Lumber Company*, 200 U.S. 321 (1906). The case is a standard reference to warn attorneys not to rely on the syllabus of a reported case. To cite Supreme Court cases, one must cite to the official Supreme Court reporter, United States Reports (abbreviated as U.S.).

---

[3] http://bulk.resource.org/courts.gov/c/raw/code/readme.txt

Table 3.3   Top 100 most cited cases in the dataset

Citations	Case	Parties	Citing cases	Context
518	200 U.S. 321	*United States, Appt. v. Detroit Timber & Lumber Company*	Retrieve	Show
272	428 U.S. 153	*Troy Leon Gregg. Petitioner v. State of Georgia*	Retrieve	Show
236	384 U.S. 436	*Ernesto A. Miranda. Petitioner v. State of Arizona*	Retrieve	Show
236	543 U.S. 220	*United States v. Booker*	Retrieve	Show
196	372 U.S. 335	*Clarence Earl Gideon. Petitioner v. Louie L. Wainwright*	Retrieve	Show
191	304 U.S. 64	*Erie Railroad. CO. v. Tompkins*	Retrieve	Show
187	310 U.S. 296	*Cantwell et al. v. State of Connecticut*	Retrieve	Show
185	297 U.S. 288	*Ashwander et al. v. Tennessee Valley Authority*	Retrieve	Show
168	376 U.S. 254	*The New York Times Company. Petitioner, v. L. B. Sullivan*	Retrieve	Show
167	209 U.S. 123	*Ex Parte: Edward T. Young, Petitioner*	Retrieve	Show
163	118 U.S. 356	*Yick Wo v. Hopkins, Sheriff*	Retrieve	Show
162	357 U.S. 449	*National Association for the Advancement of Colored People (NAACP) v. Patterson*	Retrieve	Show
161	371 U.S. 415	*National Association for the Advancement of Colored People (NAACP) v. Button*	Retrieve	Show
160	6S.Ct. 524	*Boyd and others. Claimants, etc., v. United States*	Retrieve	Show

In the example 200 U.S. 321, "200" is the volume of the U.S. Reports, "U.S." is the abbreviation of the Reporter, and "321" is the first page of the case in the volume. A case citation is also included in the year the case was decided as we shall see shortly.

Figure 3.11 shows a few examples of citations to the case *Miranda v. Arizona*, 384 U.S. 436 (1966). The references of the citing cases are listed in the left column. Instances of citations are highlighted in the context column. In this dataset, the case was cited subsequently by 145 cases.

The citation count is a commonly used metric to measure how often it is referred to by subsequent decisions. Another useful metric, citation burst, measures how fast the citation count increases over time. Figure 3.12 shows a visualization of cases with citation bursts based on a subset of cases that are cited along with the 1966 *Miranda v. Arizona* case. The light colored bar shows the period of time during which a case has been decided; thus, it is available in the pool to be cited. The dark colored bar indicates the

Supreme Court Case Citation Context

Miranda v. Arizona: 145 cases found:

Case	Context
384 U.S. 719	67, 202 A.2d 669 (1964). This is an adequate state ground which precludes us from testing the coerced confession claim on the present review, whatever may be the significance of the state court's reliance on its procedural rule in federal habeas corpus proceedings. See Fay v. Noia, 372 U.S. 391, 83 S.Ct. 822 (1963). 35 The judgment of the Supreme Court of New Jersey is affirmed. 36 Affirmed. 37 Mr. Justice CLARK concurs in the opinion and judgment of the Court. He adheres, however, to the views stated in his separate opinion in Miranda v. Arizona, 384 U.S. 499, 86 S.Ct. 1640. 38 Mr. Justice HARLAN, Mr. Justice STEWART, and Mr. Justice WHITE concur in the opinion and judgment of the Court. They continue to believe, however, for the reasons stated in the dissenting opinions of Mr. Justice Harlan and Mr. Justice White in Miranda v. Arizona and its companion cases, 384 U.S. 504, 526, 86 S.Ct. 1643, 1655, that the new constitutional rules promulgated in those cases are both unjustified and unwise. 39 Mr. Justice BLACK, with whom Mr. Justice DOUGLAS joins, dissents from the Court
384 U.S. 737	v. Richmond, 365 U.S. 534, 81 S.Ct. 735, 5 L.Ed.2d 760 (1961). The sole issue presented for review is whether the confessions were voluntarily given or were the result of overbearing by police authorities. Upon thorough review of the record, we have concluded that the confessions were not made freely and voluntarily but rather that Davis' will was overborne by the sustained pressures upon him. Therefore, the confessions are constitutionally inadmissible and the judgment of the court below must be reversed. 4 Had the trial in this case come after our decision in Miranda v. Arizona, 384 U.S. 436, 86 S.Ct. 1602, 16 L.Ed.2d 694, we would reverse summarily. Davis was taken into custody by Charlotte police and interrogated repeatedly over a period of 16 days. There is no indication in the record that police advised him of any of his rights until after he had confessed orally on the 16th day.1 This would be clearly improper under Miranda. Id., 384 U.S. at 478—479, 492, 86 S.Ct. at 1630, 1637. Similarly, no waiver of rights could be inferred from this record since it shows only that Davis was repeatedly interrogated and that he denied the alleged offe
384 U.S. 757	hat might be used to prosecute him for a criminal offense. He submitted only after the police officer rejected his objection and directed the physician to proceed. The officer's direction to the physician to administer the test over petitioner's objection constituted compulsion for the purposes of the privilege. The critical question, then, is whether petitioner was thus compelled 'to be a witness against himself.'6 7 If the scope of the privilege coincided with the complex of values it helps to protect, we might be obliged to conclude that the privilege was violated. in Miranda v. Arizona, 384 U.S. 436, at 460, 86 S.Ct. 1602, at 1620, 16 L.Ed.2d 694, at 715, the Court said of the interests protected by the privilege: 'All these policies point to one overriding thought: the constitutional foundation underlying the privilege is the respect a government state or federal—must accord to the dignity and integrity of its citizens. To maintain a 'fair state-individual balance,' to require the government 'to shoulder the entire load,' * * * to respect the inviolability of the human personality, our accusatory system of criminal justice demands that the governme

FIGURE 3.11   Examples of citations to the case *Miranda v. Arizona*, 384 U.S. 436 (1966).

Top 20 references with strongest citation bursts 1966–2004

References	Year	Strength	Begin	End
86 SCT 1772, 1700, 86 SCT 1772, V86, P1772	1700	6.0032	1966	1972
6 SCT 524, 1886, 6 SCT 524, V6, P524	1886	4.5558	1966	1976
384 US 719, 1966, 384 US 719, V384, P719	1966	5.6483	1967	1972
380 US 400, 1965, 380 US 400, V380, P400	1965	4.4895	1967	1973
97 SCT 1232, 1700, 97 SCT 1232, V97, P1232	1700	5.2428	1977	1989
99 SCT 2248, 1700, 97 SCT 2248, V99, P2248	1700	5.407	1979	1987
442 US 707, 1979, 442 US 707, V442, P707	1979	7.1611	1980	1990
99 SCT 2560, 1700, 99 SCT 2560, V99, P2560	1700	6.7932	1980	1990
451 US 477, 1981, 451 US 477, V451, P477	1981	7.3863	1981	1991
446 US 291, 1980, 446 US 291, V446, P291	1980	5.6682	1981	1991
451 US 454, 1981, 451 US 454, V451, P454	1981	4.9238	1982	1992
99 SCT 2568, 1700, 99 SCT 2568, V99, P2568	1700	4.8628	1984	1990
465 US 638, 1984, 465 US 638, V465, P638	1984	4.7568	1984	1991
462 US 1039, 1983, 462 US 1039, V462, P1039	1983	4.1682	1984	1990
467 US 649, 1984, 467 US 649, V467, P649	1984	4.9726	1985	2004
470 US 298, 1985, 470 US 298, V470, P298	1985	7.3359	1986	2004
475 US 412, 1986, 475 US 412, V475, P412	1986	6.5998	1986	2004
468 US 420, 1984, 468 US 420, V468, P420	1984	5.1291	1986	2004
475 US 625, 1986, 475 US 625, V475, P625	1986	4.5232	1987	2004
489 US 288, 1989, 489 US 288, V489, P288	1989	4.8128	1990	2004

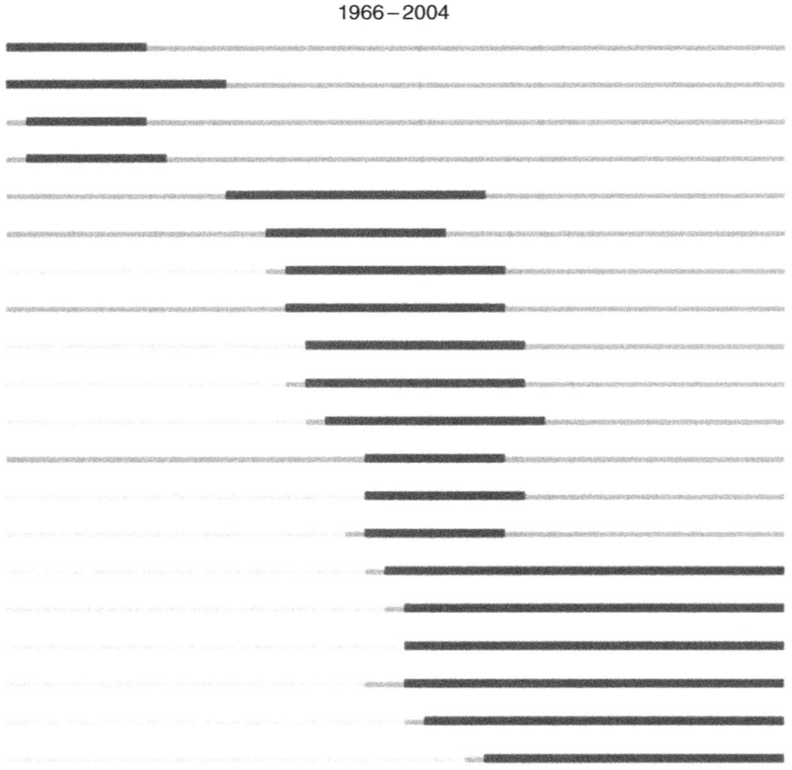

FIGURE 3.12 Top 20 cases with the strongest citation bursts.

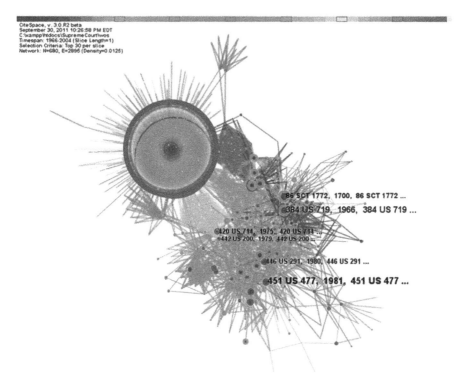

CiteSpace, v. 3.0.R2 beta
September 30, 2011 10:26:58 PM EDT
C:\xampp\htdocs\SupremeCourt\wos
Timespan: 1966-2004 (Slice Length=1)
Selection Criteria: Top 30 per slice
Network: N=680, E=2895 (Density=0.0125)

86 SCT 1772, 1700, 86 SCT 1772 ...
384 US 719, 1966, 384 US 719 ...
420 US 714, 1975, 420 US 714 ...
442 US 200, 1979, 442 US 200 ...
446 US 291, 1980, 446 US 291 ...
451 US 477, 1981, 451 US 477 ...

FIGURE 3.13   A visualization of a cocitation network of cases cited with the case *Miranda v. Arizona*, 384 U.S. 436 (1966). The size of a circle is proportional to the citations received by the case. Red circles represent cases with citation bursts. (*See insert for color representation of the figure.*)

duration in which a burst of citation is detected. Since these cases are cited with the 1966 *Miranda v. Arizona* case, the year of citation cannot be earlier than 1966. The thinner light colored bar shows the time before the case was decided. With a comprehensive dataset, this method can be used to identify cases with abruptly increased citations.

If two cases, *a* and *b*, are cited by another case, *a* and *b* are cocited. The number of times they are cocited is called cocitation count. Cocitations are commonly used to form a cocitation network because nodes in the network are connected if they are cocited. Cocitation networks provide a useful representation for the analysis of interrelationships between individual entities and trends at a higher level of aggregation. In later chapters, we will introduce the underlying techniques in more detail. Figure 3.13 shows a visualization of a cocitation network of cases cited along with the case *Miranda v. Arizona*, 384 U.S. 436 (1966). The size of a circle represents the citations to a corresponding case. Red circles indicate cases with citation bursts, for example, 451 U.S. 477 (1981).

In order to analyze the Supreme Court Opinions further, we need to develop techniques that can help us to identify patterns associated with a complex adaptive system, which we will address in the next few chapters. We will demonstrate how to answer some of the relatively straightforward questions. For example, for a given case, what are the precedent cases that are similar in one or more aspects? More interestingly, we will address the need to identify information that plays critical roles in the development of a complex adaptive system such as identifying early signs for revolutionary changes in scientific knowledge.

3.5   APPLE VERSUS SAMSUNG

Apple and Samsung have been fighting over patent infringement cases even as their products are flying off the shelf. What is patentable? What is supposed to be protected? This is another area where arguments are made with different interpretations of evidence.

Besides high-profile patent infringement cases, there are also many smaller-scale cases over patents or litigations. A large portion of such lawsuits are filed by companies known as patent trolls. Trolls are companies that make their living by suing others, but do not produce their own products. Patent trolls make their money by licensing or litigating others' intellectual properties. A recently published 110-page research report authored by Robin Feldman, Thomas Ewing, and Sara Jeruss studied patent litigations filed across four years, 2007–2008 and 2011–2012, and analyzed about 13,000 cases and 30,000 patents involved. They found that as of 2012, 58.7% of patent lawsuits were filed by patent trolls, which represents a sharp increase from 24.6% in 2007. New companies are emerging to deal with the increasing demand in finding relevant information quickly and effectively. Lex Machina as Stanford University's spinoff company[4] provides legal analytics to companies and law firms. The company creates datasets of rich and federated information about judges, lawyers, and patents to help its clients to deal with legal disputes and patent litigations.

In the United States, patentability is defined in the Manual of Patent Examining Procedure (MPEP), published by the USPTO and used by patent examiners and patent lawyers. USPTO grants patents on a presumptive basis, which means that patentability is acceptable in the absence of evidence that supports why an idea should not have been granted. In light of new evidence, this could change patentability, usually in the court of law. According to the European Patent Office,[5] patentability has four basic requirements:

1. There must be an "invention," belonging to any field of technology.
2. The invention must be "susceptible of industrial application."
3. The invention must be "new."
4. The invention must involve an inventive step.

The European Patent Convention (EPC) does not consider the following as patentable inventions: discoveries, scientific theories, and mathematical methods; aesthetic creations; schemes, rules, and methods for performing mental acts; playing games or doing business; and computer programs.

---

[4] https://lexmachina.com/about/
[5] http://www.epo.org/law-practice/legal-texts/html/guidelines/e/g_i.htm

On October 7, 2012, *The New York Times* published "Fighters in a Patent War,[6]" which detailed the patent lawsuits over smartphone patents. It was revealed that Samsung infringed the patented one-finger scrolling design for devices such as the Galaxy.

Technological giants such as Apple, Google, and Microsoft are all deeply involved in the patent war. Google bought Motorola Mobility in 2011 for US $12.5 billion, largely for its patent portfolio.[7]

3.6 SUMMARY

John H. Wigmore's chart method was first known in 1913 as an analytic and systematic way to marshal evidence and estimate probative force credentials of evidence. In essence, his method is a way of constructing and establishing the credibility of an argument. The Wigmore Chart, however, was criticized for being too complicated for lawyers and members of the public. It was widely regarded as a failed invention until the major work of Kadane and Schum began to integrate Wigmore's ideas with modern versions of belief networks—Bayesian inference networks.

History has highlighted the importance in considering competing hypotheses and seemingly conflicting evidence simultaneously. In 2013, the Nobel Prize in Economics was awarded to three economists who have held competing theories.

In this chapter, a recurring theme of these examples is that the same evidence is often subject to an array of diverse and even conflicting interpretations. The same evidence may be viewed in a different context from a different perspective. A strong argument is the one that can accommodate alternative and distinct views.

The notion of a gap is also recurring along with a generalized optimization through an iterative process of evolution. The analysis of the reviews of *The Da Vinci Code* underlines the value of explanatory approaches to the understanding of gaps. The brief introduction to case citations in Supreme Court Opinions is simply to raise some questions to be addressed in the next few chapters.

[6] http://www.nytimes.com/interactive/2012/10/08/business/Fighters-in-a-Patent-War.html
[7] http://www.nytimes.com/2012/10/08/technology/patent-wars-among-tech-giants-can-stifle-competition.html

BIBLIOGRAPHY

Algranati, D. J., & Kadane, J. B. (2004). Extracting Confidential Information from Public Documents: The 2000 Department of Justice Report on the Federal Use of the Death Penalty in the United States. *Journal of Official Statistics*, *20*, 97–113.

Anderson, T., Schum, D., & Twining, W. (2005). *Analysis of evidence* (2nd ed.). Cambridge, UK: Cambridge University Press.

Bernanke, B. S. (2013, November 8). The crisis as a classic financial panic. *The 14th Jacques Polak annual research conference*, Washington, DC. Retrieved June 1, 2013, from http://www.federalreserve.gov/newsevents/speech/bernanke20131108a.htm. Accessed on March 7, 2014.

Boston Public Library Staff. (1979). *Sacco-Vanzetti: Developments and reconsiderations*. Boston: Boston Public Library.

Chang, C.-C., & Lin, C.-J. (2001). *LIBSVM: A library for support vector machines*. Retrieved June 1, 2013, from http://www.csie.ntu.edu.tw/~cjlin/libsvm. Accessed on March 7, 2014.

Chen, C., Ibekwe-SanJuan, F., SanJuan, E., & Weaver, C. (2006, October). *Visual analysis of conflicting opinions*. Paper presented at the Proceedings of the IEEE Symposium on Visual Analytics Science and Technology (VAST), Baltimore, MA.

Dawid, A. P., & Evett, I. W. (1997). Using a graphical method to assist the evaluation of complicated patterns of evidence. *Journal of Forensic Sciences*, *42*, 226–231.

Dawid, A. P., Leucari, V., & Schum, D. A. (2005). *Analysis of complex patterns of evidence in legal cases: Wigmore charts vs. Bayesian networks*. Retrieved June 1, 2013, from http://128.40.111.250/evidence/content/leucariA1.pdf. Accessed on March 7, 2014.

Edwards, W. (1991). Influence diagrams, Bayesian imperialism, and the Collins case: An appeal to reason. *Cardozo Law Review*, *13*, 1025–1074.

Felix, D. (1965). *Protest: Sacco-Vanzetti and the Intellectuals*. Bloomington, IL: Indiana University Press.

Goodwin, J. (2000). Wigmore's chart method. *Informal Logic*, *20*(3), 223–243.

Grossman, J. (1962, January). The Sacco-Vanzetti case reconsidered. *Commentary*, 31–44.

Kadane, J. B., & Schum, D. (1992). Opinions in dispute: The Sacco-Vanzetti case. In: J. M. Bernardo, J. O. Berger, A. P. Dawid, & A. F. M. Smith, Eds. *Bayesian statistics 4* (pp. 267–287). Oxford, UK: Oxford University Press.

Kadane, J. B., & Schum, D. A. (1996). *A probabilistic analysis of the Sacco and Vanzetti evidence*. New York: Wiley. Retrieved June 1, 2013, from http://www.amazon.com/Probabilistic-Analysis-Vanzetti-Probability-Statistics/dp/0471141828; http://books.google.com/books?id=t7QjVbn825wC&printsec=frontcover#v=onepage&q&f=false. Accessed on March 7, 2014.

Lazarowitz, E. (2013, January 18). Bernanke missed signs of crises. *New York Daily News*. Retrieved June 1, 2013, from http://www.nydailynews.com/news/national/bernanke-missed-signs-crisis-article-1.1243044?print. Accessed on March 7, 2014.

Linder, D. (2000). Summary of evidence in the Sacco & Vanzetti case. Retrieved June 1, 2013, from http://law2.umkc.edu/faculty/projects/ftrials/SaccoV/s&vevidence.html. Accessed on March 7, 2014.

Pang, B., & Lee, L. (2004). *A sentimental education: Sentiment analysis using subjectivity summarization based on minimum cuts* (pp. 271–278). Proceedings of the 42nd annual meeting on Association for Computational Linguistics, July 21–26, 2004. Stroudsburg, PA.

Robin, F., Thomas, E., & Sara, J. (2013). *The AIA 500 expanded: Effects of patent monetization entities*. Retrieved June 1, 2013, from http://papers.ssrn.com/sol3/papers.cfm?abstract_id=2247195. Accessed on March 7, 2014.

Russell, F. (1962). *Tragedy in Dedham: The story of the Sacco-Vanzetti case*. New York: McGraw-Hill.

Tillers, P. (2004, 2005). *Picturing factual inference in legal settings*. http://tillers.net/pictures/picturing.html. Accessed on November 16, 2013.

UCL. Evidence, inference and enquiry: Towards an integrated science of evidence. *Enquiry, evidence and facts: An interdisciplinary conference*. Retrieved June 1, 2013, from http://128.40.111.250/evidence/. Accessed November 21, 2013.

CHAPTER 4

Visualizing the Growth of Knowledge

Scientists and scholars have learned, most likely before the beginning of their career, why they need to share their work widely and persistently through presentations at conferences and publications in journals and books. All these publications form the literature, which contains not only valuable clues, inspirational trails, and challenging problems, but also, inevitably, noises and even deceptive spots. The literature as a whole contains descriptions, hypotheses, findings, criticisms, and speculations resulting from numerous mental models of scientists, scholars, thinkers, and practitioners. As such, its greatest value is that it grows organically as new publications flow in and as old and unnoticed publications fade out. On the other hand, the amount of information grows much faster than the rate we can possibly digest, transform, and integrate with our own mental models. A profound challenge to its users and contributors is how to cut to the chase and focus on the most important information that the literature can offer.

Shortly after the terrorist attacks on September 11, 2001, Dr. Sandro Galea and his colleagues conducted a study to investigate the prevalence of post-traumatic stress disorder (PTSD) in the general population of New York City. The study consisted of three random dialing telephone surveys of adults in the metropolitan area. The three surveys were conducted 1, 4, and 6 months after September 11, 2001, with 1008, 2001, and 2752 participants, respectively. The prevalence of probable PTSD dropped from 7.5% to 0.6% during the period of the study. Galea and his team noticed that although the prevalence was consistently higher among persons who were directly affected by the attacks, a substantial number of persons who were not directly affected by the attacks, surprisingly, also met criteria for probable PTSD. The surprising findings in people who were not directly affected by traumatic events sent a signal that was inconsistent with what was generally believed by PTSD experts. In the past, it was believed that PTSD was possible if one experienced a trauma directly. The first part of Galea's study was published in the *New England Journal of Medicine* on March 28, 2002. According to Google Scholar, as of December 22, 2013,

The Fitness of Information: Quantitative Assessments of Critical Evidence, First Edition. Chaomei Chen.
© 2014 John Wiley & Sons, Inc. Published 2014 by John Wiley & Sons, Inc.

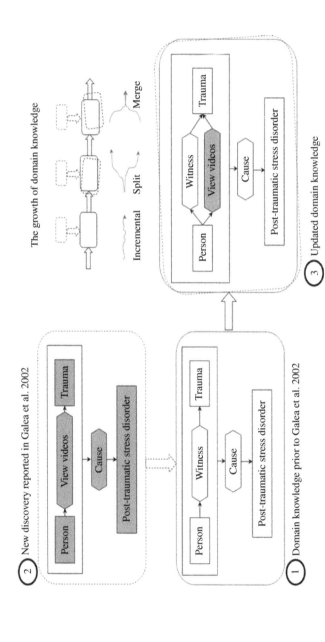

FIGURE 4.1 An ultimate ability to reduce the vast volume of scientific knowledge in the past and a stream of new knowledge to a clear and precise representation of a conceptual structure: ① domain knowledge prior to Galea et al. (2002), ② new discovery reported in Galea et al. (2002), and ③ updated domain knowledge.

it was cited 1287 times. The full report of the study, published in the *American Journal of Epidemiology* in 2003, was cited 394 times.

Galea's study has drawn researchers' attention to a no-man's land, at least to an extent. The prevalence among people who had experienced the terrorist attacks indirectly was largely due to extensive graphical news coverage on mass media. Galea's study revealed an invisible gap between their new finding and the contemporary literature.

The ability to grasp the significance of the surprising finding and to bridge the gap accordingly relies on a good understanding of the study in a much broader and complex context. The six-page article itself is the tip of an iceberg. A substantial amount of time and effort is typically required to identify the significance of a single publication. Much more domain knowledge is required to be able to see how a new study is related to the body of existing knowledge.

Figure 4.1 illustrates the significance of Galea's study in terms how it alters the mental model of PTSD. Reducing the complexity of the literature to this level of simplicity requires a considerable level of domain expertise. On the other hand, in order to cut to the chase when facing the rapidly increased volume of new information, what can we do differently, at least to supplement the traditional systematic reviews of what has been done and what is attracting everyone's attention?

4.1 PROGRESSIVE KNOWLEDGE DOMAIN VISUALIZATION

Knowledge does not have clear-cut boundaries in its own right. We often talk about the body of knowledge, a patch of knowledge, or the universe of knowledge with an implicit reference to an underlying topic or phenomenon. Do all the publications mentioning the word *trauma* collectively cover the entire knowledge behind the word? Do all the publications in journals that have the word *trauma* in their titles adequately represent what is known about the topic? The relevance of an article to a topic falls on a continuum rather than a discrete set of points. There does not seem to be a natural and objective way to draw a line to separate what is absolutely relevant and what is definitely not. Rather, the separation is subtle, gradual, and fractal.

Instead of using terms such as area, field, or discipline to define the scope of relevant knowledge, we use the term *domain* to refer to a domain of knowledge. The term *domain* is used to underline the ambiguity and uncertainty of its boundary. The notion of a knowledge domain provides a more generic, flexible, and accurate reference to the organization of knowledge than terms such as a topic area of research, a field of study, a discipline, or a combination of any of these entities.

The goal of progressive visualization of a domain of knowledge is to reveal the structure of the organically defined domain and depict how it evolves over time. The definition is intentionally broad. Just like the subjectivity of evidence as we have seen in Chapter 3, a knowledge domain is defined by the perspective we choose to take!

Our mental models determine what may be considered as part of a knowledge domain. A domain of knowledge usually deals with many topics that may appear to be loosely related unless they are seen from a unifying perspective. An existing domain appears to be relatively stable merely because we have been used to the same perspective. A new domain may come into being because a creative perspective is found. Such perspectives may be inspired by the external world as well as by our internal world. The notion of a knowledge domain is broader than a paradigm in that a single knowledge domain may accommodate multiple competing paradigms. We use the term *knowledge domain* to emphasize the dynamic nature of the phenomenon.

4.1.1   The Structure of a Knowledge Domain

The idea of a paradigm shift from Thomas Kuhn's *Structure of Scientific Revolutions* is widely known. In Kuhn's theory, science evolves by repeatedly going through a series of states, namely, establishing a new paradigm, expanding and consolidating the paradigm, the paradigm in crisis, and a revolution—the paradigm is being replaced by a new paradigm. Kuhn's work has generated deep interests in interpreting major events in science using this framework.

The rich and widely accessible literature of scientific disciplines has attracted a lot of attention as one of the major sources of input for the study of the structure and dynamics of paradigms. As Kuhn pointed out, the writings of scientists in the relevant literature contain valuable trails of the dynamics of competing paradigms as they emerge, transform, and disappear.

A particularly valuable source of clues of scientific knowledge is how scientists articulate their work in terms of how it is related to prior contributions of their peers. In particular, scientists make explicit references to previous publications in the literature. Such instances of references are called *citations*. For example, the Galea et al.'s 2002 article on the prevalence of PTSD in New York City cited 32 previously published articles. Their article was in turn cited by 1287 documents on the Internet according to Google Scholar. The study of patterns of citations is called *citation analysis*, which is part of bibliometrics, which in turn belongs to scientometrics.

The foundation of citation analysis was laid down in the 1950s by Eugene Garfield. He created the Science Citation Index (SCI) and the Social Sciences Citation Index (SSCI) to enable citation indexing, which is an alternative way to the traditional information retrieval approaches. In traditional retrieval approaches, the relevance of a document to a search is in essence determined based on the similarity between the document and the search query in terms of the usage of words. However, if two documents use completely different vocabularies, then the traditional approaches will break down due to the infinitely wide gap between them even if the two documents may be still semantically relevant. The idea of citation indexing was proposed to provide an alternative way to detect possible connections between the two documents. If two documents cite a common set of references, then they are likely to have something in common. In terms of Bayes's rules, the probability that the two documents are semantically similar given the fact that they share some common references is P(two documents are similar | they cite common references). It can be estimated from three other probabilities, namely, P(they cite common references | they are similar), P(two documents cite common references), and P(two documents are relevant). Kessler's bibliographic coupling was one of the earliest methods proposed to calculate similarities based on overlapping references (Kessler, 1963).

Another way to estimate the citation-based similarity between two articles was proposed by Henry Small (1977). His method has been adopted by many researchers for cocitation analysis. The similarity between two articles is determined by how frequently the two articles were cited together in the past. The cocitation analysis differs from the bibliographic coupling method in several ways. The focus of a cocitation analysis is on the set of references that have been cited at least once. Articles that have no citations at all will not appear in the results of a cocitation analysis. In contrast, the bibliographic coupling method does not require a minimum number of citations. Results of a bibliographic coupling analysis in general include more recently published articles than that of a cocitation analysis. Both methods represent different perspectives toward patterns of citations. It is a good idea to consider both of them when analyzing a knowledge domain.

Pairwise similarities between scientific publications make it possible to form a network of scientific publications. Such networks can be divided into subnetworks based on the strengths of similarity links. Each subnetwork corresponds to a component at a higher level of granularity. A subnetwork formed by a group of similar publications on PTSD, for example, can be seen as a composite node of the topic PTSD. Such subnetworks may also represent entities at even higher levels of abstraction, namely, paradigms and disciplines.

The decomposition of a network may change over time. Two components may merge to become one. One component may split into multiple components. The dynamics of the underlying network's structure reflects the evolution of scientific literature, which is in turn driven by the dynamics of scientific knowledge. Henry Small revealed in his pioneering cocitation study how the landscape of collagen research changed from one year to another. Clusters of cocited references were depicted by contour lines. An existing cluster in one year may grow larger in the next year or become diminished. A new cluster may suddenly emerge and dominate the landscape. Small's study predated many modern visualization techniques, but the language in his description was vivid enough for everyone to picture the changes.

Using the metaphor of a landscape to portrait an abstract space is an attractive choice. We are all familiar with natural landscapes that often consist of distinct features such as mountains, hills, lakes, and rivers. The distribution of these features makes each view of a landscape unique. The metaphor naturally implies how we may be able to engage with the landscape and what actions would be appropriate. We may walk along a river, climb up a hill, or fly over the landscape. In this way, we may realize how we are supposed to read a landscape of a scientific field and how to interpret the height of a hill of cited references and the distance between the peaks of two disparate topics. Widely known examples of such design include the ThemeView from the Pacific Northwest National Laboratory in mid 1990s.

Animated visualization techniques provide further options for us to highlight the evolution of such a landscape over the years. It may not be easy to spot subtle changes

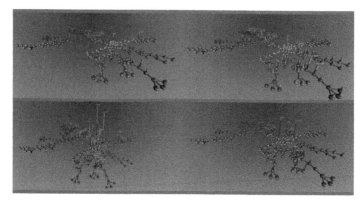

FIGURE 4.2   An animated visualization of a network-based landscape view of the literature on mad cow disease. (*See insert for color representation of the figure.*)

shown in visualizations in consecutive years in still images, whereas animated visualizations can make it clear how these changes transform over time. In our earlier work (Chen & Kuljis, 2003), we animated visualizations of the landscape of scientific revolutions in physics and medicine to uncover holistic patterns of the development of a research field. Figure 4.2 is a frame of an animated visualization of the research literature on mad cow disease. The connection between two spheres represents the strength of their cocitation. The height of the vertical bar on top of a sphere represents the citations it received. The colors represent major components identified by principal component analysis (PCA).

We realized from the analysis of animated visualizations that although these models may faithfully replicate the actual growth of scientific literature, potentially insightful information tends to be overwhelmed by less critical information and common patterns. It becomes clear that it is necessary to make the important information standout. Progressive knowledge domain visualization was developed to improve the capability of these techniques so that groundbreaking articles can be characterized by distinguishable visual features (Chen, 2004).

4.1.2   Research Fronts and Intellectual Bases

In order to capture the structure and dynamics of a knowledge domain from the literature, we incorporate a wide range of principles from philosophy of science, sociology of science, and a few other relevant fields. Basic concepts associated with these principles are introduced as follows.

A research front represents the state of the art of a field. It represents the mental model of the scientific community. A research front is defined by the collective content of all the relevant publications contributed by the scientific community. The collection of all the references cited by the scientific community serves as the intellectual base of a research front. The connection between a research front and its intellectual base is established in terms of citations between them. Figure 4.3 illustrates this scenario.

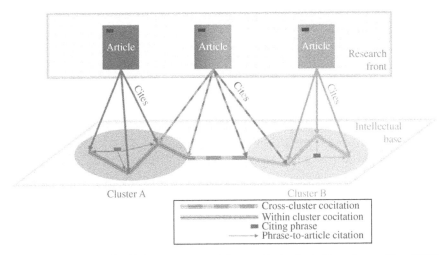

FIGURE 4.3   The relationship between a research front and its intellectual base. Source: Chen (2006).

The knowledge domain visualization has focused on three primary tasks for understanding a body of scientific literature or other types of documents such as court decision reports and patent applications:

1. Improving the clarity of important patterns in individual networks
2. Characterizing the transformation of one network to next
3. Identifying critical nodes and paths across consecutive networks

One of the commonly acknowledged aesthetic criteria in visualizing a complex network is that the number of crossing links should be minimized because they sidetrack attention to crosses that have no intended meaning. The visualization of a network that minimizes the artifact of the underlying drawing algorithm is not only aesthetically pleasing but also more efficient to guide our attention to concentrate on the most significant patterns. A common practice is to filter out less significant links from a network and retain the ones that are essential to convey structural properties of the network. Algorithms such as minimum spanning trees (MSTs) and Pathfinder network scaling are commonly used link reduction options.

Pathfinder network scaling is originally developed by cognitive scientists to build procedural models based on subjective ratings (Schvaneveldt, 1990). It uses a more sophisticated link elimination mechanism than an MST algorithm. It retains the most important links and preserves the integrity of the network. Every network has a unique Pathfinder network, which contains all the alternative MSTs of the original network.

Pathfinder network scaling aims to simplify a dense network while preserving its salient properties. The topology of a Pathfinder network is determined by two parameters r and q. The r parameter defines a metric space over a given network based on the Minkowski distance so that one can measure the length of a path connecting two nodes in the network. The Minkowski distance becomes the familiar Euclidean distance when $r=2$. When $r=\infty$, the weight of a path is defined as the maximum weight of its component links, and the distance is known as the maximum value distance. Given a metric space, a triangle inequality can be defined as follows:

$$w_{ij} \leq \left(\sum_k w^r n_k n_{k+1} \right)^{1/r}$$

where w_{ij} is the weight of a direct path between i and j, and $wn_k n_{k+1}$ is the weight of a path between n_k and n_{k+1}, for $k=1,2,...,m$. In particular, $i = n_1$ and $j = n_k$. In other words, the alternative path between i and j may go all the way round through nodes $n_1, n_2, ..., n_k$ as long as each intermediate link belongs to the network.

If w_{ij} is greater than the weight of alternative path, then the direct path between i and j violates the inequality condition. Consequently, the link $i-j$ will be removed because it is assumed that such links do not represent the most salient aspects of the association between the nodes i and j.

The q parameter specifies the maximum number of links that alternative paths can have for the triangle inequality test. The value of q can be set to any integer between 2 and $N-1$, where N is the number of nodes in the network. If an alternative path has a lower cost than the direct path, the direct path will be removed. In this way, Pathfinder reduces the number of links from the original network, while all the nodes remain untouched. The resultant network is also known as a minimum-cost network.

The strength of Pathfinder network scaling is its ability to derive more accurate local structures than other comparable algorithms such as multidimensional scaling (MDS) and MST. However, the Pathfinder algorithm is computationally expensive. The maximum pruning power of Pathfinder is achievable with $q=N-1$ and $r=\infty$; not surprisingly, this is also the most expensive one because all the possible paths must be examined for each link. Some recent implementations of Pathfinder networks reported the use of the set union of MSTs.

By characterizing the transformation of a network in consecutive timeframes, we can find what happened to a previously predominant cluster of cocited references and how a newly emerged cluster was related to the rest of the landscape. Intellectual structures of a knowledge domain before and after a major conceptual revolution can be fundamentally different as new theories and new evidence become predominant. Cocitation networks of citation classics in a field may differ from cocitation networks of newly published articles. The key question is: "what is the most informative way to merge potentially diverse networks?"

A merged network needs to capture the important changes over time in a knowledge domain's cocitation structure. We need to find when and where the most influential changes took place so that the evolution of the domain can be characterized and visualized. Few studies in the literature investigated network synthesis from a domain-centric perspective. The central idea of our new method is to visualize how different network representations of an underlying phenomenon can be informatively stitched together.

Visually salient features in a landscape provide information on critical paths and turning points that triggered fundamental changes in the knowledge of science.

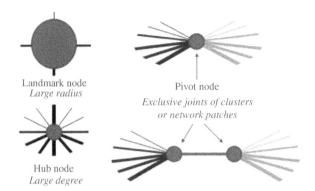

FIGURE 4.4 Three types of salient nodes in a cocitation network. Source: Chen (2004).

The goal of our visual and computational approach is to draw users' attention to these important features so that we can advance our understanding of the evolution of scientific domains.

The importance of a node in a cocitation network can be identified by the local topological structure of the node and by additional attributes of the node. We are particularly interested in three types of nodes: (1) landmark nodes, (2) hub nodes, and (3) pivot nodes (see Figure 4.4).

A landmark node is a node that has extraordinary attribute values. For example, a highly cited article tends to provide an important landmark regardless of how it is cocited with other articles. Landmark nodes can be rendered by distinctive visual–spatial attributes such as size, height, or volume. A hub node has a relatively large node degree. A widely cocited article is a good candidate for significant intellectual contributions. A high-degree hub-like node is also easy to recognize in a visualized network. Both landmark nodes and hub nodes are commonly used in network visualization. Although the concept of pivot nodes is available in various contexts, the way they are used in our method is novel. Pivot nodes are joints between different networks. They are either the common nodes shared by two networks or the gateway nodes that are connected by internetwork links. Pivot nodes have an essential role in our method.

4.1.3 Strategies of Scientific Discoveries

Predicting future citations has been a recurring topic of interest for researchers, evaluators, and science policymakers. The predictive power of a diverse range of variables has been tested in the literature. As shown in Table 4.1, most of the commonly studied variables can be categorized into a few groups according to their parent classes where they belong to. For example, the number of pages of a paper is an attribute of the paper as an article. The number of authors of a paper is an attribute of the authorship of the paper. One can expect that even more variables will be added to the list. We expect to demonstrate that our theory of transformative discovery provides a theoretical framework to accommodate this diverse set of attributive variables and a consistent explanation for most of them.

Table 4.1   Variables associated with articles that may be predictive of their subsequent citations

Components	Attributive variables	Hypotheses derived from the theory of discovery
Article	Number of pages	Boundary spanning needs more text to describe.
	Number of years since publication	
Authorship	Number of authors	More authors are more likely to contribute from diverse perspectives.
	Reputation (citations, *h*-index)	
	Gender	
	Age	
	Position of last name in alphabet	
Impact	Citation counts	The value of the work is recognized.
Usage	Download times	The value of the work is recognized.
Abstract	Number of words	Transformative ideas tend to be more complex than monolith ones. More words needed to express more complex ideas.
	Structured (yes/no)	
Content	Type of contributions: tools, reviews, methods, data, etc.	
	Scientific rigorous of study design	
Reference	Number of references	More references needed to cover multiple topics that are being synthesized.
Discipline	Number of disciplines	It is more likely that the work synthesizes multiple disciplines.
Country	Number of countries	It is more likely that authors from different countries bring in distinct perspectives.
Institution	Number of institutions	It is more likely that authors from different institutions bring in distinct perspectives.
Journal	Impact factor	
	Indexed by different databases	
Sponsored	Yes/no	

Several studies focus on the relationship between earlier download times and subsequent citations (Brody, Harnad, & Carr, 2006; Lokker, McKibbon, McKinlay, Wilczynski, & Haynes, 2008; Perneger, 2004). Perneger (2004) studied 153 papers published in one volume of the journal *BMJ* in 1999 (volume 318) along with the full paper download times within the first week of publication and their citations as of May 2004 recorded in the Web of Science. Perneger coded each paper in terms of its study design using seven categories, namely, randomized trials, systematic reviews, prospective studies, case–control studies, cross sectional surveys, qualitative studies, and other designs. He found a statistically significant positive Pearson correlation of

0.50 ($p<0.001$) between citations and the download times within the first week. He also found that 33% of variance can be explained by hits (download times) and the length of a paper. A correlation of 0.4 was found between citations and downloads of articles in the e-print archive repository arXiv (Brody et al., 2006), although the amount of variance explained (16%) was relatively low.

In a more recent analysis, a group of researchers at McMaster University, Canada, studied whether 20 article and journal variables can predict citations of 1274 articles from 105 journals published between January and June 2005 (Lokker et al., 2008). The 20 variables include ratings of clinical relevance and newsworthiness, which are routinely collected by the McMaster online rating of evidence system. The dataset was split by 60:40 for derivation and validation. Their study shows that a multiple regression model accounted for 60% of the variance in the derivation portion of the dataset. The same model accounted for 56% of the variance in the validation dataset. Higher citations were predicted by the number of indexing services, the number of authors, cited references, clinical relevance scores, original papers, multi-center studies, and a few other variables.

Dalen and Henkens (2005) studied 1371 articles published between 1990 and 1992 in 17 demography journals in order to identify explanatory factors that may influence the visibility of an article. In particular, they were interested in whether the reputations of authors and journals had anything to do with the citations these papers would receive later on and how soon they would get their first citation. An author's reputation was estimated based on the citations of the author in 1990, the first year of the period. If an article has multiple authors, the most prominent author's reputation was used. The reputation of a journal was represented by its impact factor in 1990. They used duration analysis, originated in survival analysis, to analyze the data. The central question is: what determines the probability of an article changing from the initial state of not being cited to a state in which it is cited? In survival analysis, the role of a hazard function is to estimate the probability of transitions from the initial state. The simplest form of a hazard function is constant with no memory of how long the initial state lasts. In other words, the probability of an article moving away from the initial state in the next timeframe does not depend on how much time it has been spent in the initial state. More realistic hazard functions include positive and negative duration dependence. Positive duration dependence specifies that the longer an article has been in the initial state, that is, not being cited, the chance it will be cited improves. In contrast, negative duration dependence assumes the opposite. Dalen and Henkens chose their hazard function based on the Gompertz distribution.

The Gompertz distribution is a theoretical distribution of survival times. It was proposed by Gompertz in 1825 to model human mortality. The resultant hazard function is r.

$$y(t) = ae^{be^{ct}} \tag{4.1}$$

where a is the upper asymptote, that is, the value of $y(t \to \infty)$ in the infinite future time, b is the x displacement, c is the growth rate, and e is Euler's number. The Gompertz function models the slow growth at the initial and final stages and faster growth in intermediate stages. It has been used to model the growth of tumors, the uptake of mobile phones, and the mortality of population.

Dalen and Henkens divided articles into four categories and then used a statistical method called multinomial logit to test how explanatory factors such as authors' and journals' reputations could explain the citation patterns.

1. Articles with citations too little and/or too late (forgotten ones).
2. Articles with late citations (sleeping beauties).
3. Articles with early citations but fading off quickly (flash-in-the-pans).
4. Articles with early citation and many subsequent cites (normal science).

$$\text{Prob (Article = sleeping beauty)} = \frac{\exp(X\beta^2)}{[1+\exp(X\beta^2)+\exp(X\beta^3)+\exp(X\beta^4)]} \quad (4.2)$$

Their model shows statistically significant effects of several explanatory variables such as author reputation, the number of pages, and journal reputation (impact factor) at $p<0.01$.

The survival model of the timing of first citation identified the major role of the communication process in speeding up the uptake of a scientific paper, namely, visibility, language, and reputation of authors and journals. When the effect of a journal's quality such as the reputation of the editors and editorial policy is controlled, the duration analysis reveals the reputation effect of authors. The effect of journals becomes clear.

Dalen and Henkens's duration study essentially tells us that the reputation of the authors of an article and the reputation of the journal in which the article is published are the most critical factors for the article to gain visibility and get cited. Are we attracted by other signals? What about structural, temporal, and semantic properties of the underlying topic?

What is the extent to which quantitative rankings of highly cited authors confirm or, even more ambitiously, predict Nobel Prize awards? Between 1977 and 1992, Garfield published a series of studies of Nobel Prize winners' publications and their citations and made predictions of future Nobel Prize laureates based on existing citation data. He reported that eight Nobel laureates were found on a list of 100 most cited authors from 1981 through 1990 (Garfield & Welljamsdorof, 1992). Others on the list were seen as potential Nobel Prize winners in the future. On the other hand, it was noted that the undifferentiated rankings of the most cited authors in a given period of time could be further fine-tuned to increase the accuracy of its coverage of Nobel Prize awards. For example, the Nobel Committee sometimes selects relatively small specialties. Further dividing the list according to specialties shows that Nobel laureates in relatively small specialties are among the most cited authors in their specialties.

Methods papers of Nobel Prize winners tend to attract a disproportionably high amount of citations. More recent examples of methodological contributions include the 2007 Nobel Prize for the British embryonic stem cell research architect Martin

Evans. Garfield coined the phenomenon the *Lowry phenomenon*, referring to the classic example of Oliver Lowry's 1951 methods paper, which was cited 205,000 times up to 1990.

Research has shown that citation frequency has a low predictive power for Nobel awards because there are so many other scientists with the same or even higher citations as the few Nobel Prize winners. The greatest value of counting citations is its simplicity. Subsequent attempts to improve the accuracy of the method tend to lose the simplicity. Hirsch's *h*-index has drawn much interest also because of its simplicity despite its known limitations (Hirsch, 2005). Antonakis and Lalive intended to capture both the quality and productivity of a scholar with a new index – the index of quality and productivity (IQp) (Antonakis & Lalive, 2008). They compared the new index of Nobel winners in physics, chemistry, medicine, and economics. It is worth noting here that one should always be cautious when using quantitative indicators in qualitative decisions. The authors found that about two-thirds of Nobel winners have an IQp over 60. The authors showed that in several examples, IQp differentiated Nobel class and others more accurately than the *h*-index, including physicist Ed Witten ($h = 115$ and $IQp = 230$) and others, who have high *h*-index but relatively low IQp index, S. H. Snyder ($h = 198$, $IQp = 117$), and R. C. Gallo ($h = 155$, $IQp = 75$).

The ability to think creatively and look at a problem from a fresh perspective is known to be essential for scientific discoveries. Dunbar compared hypothesis generation strategies based on a Nobel Prize–winning discovery (Dunbar, 1993). He found that subjects performed better if they were encouraged to consider novel alternative hypotheses. A longitudinal study of a group of highly creative scientists in nanoscience and technology found that communicating effectively with a diverse group of peer scientists seems to make them more creative (Heinze & Bauer, 2007).

Philosophers of science (Laudan et al., 1986) argued that it would be useful to compare rival theories of scientific change against the history of science and see which one makes the most sense. Laudan et al. suggested that propositions of philosophical theories should be organized in such a way so that these theories can be compared in terms of individual propositions. For example, Lakatos's research program, Laudan's research tradition, and Kuhn's paradigm theory can be compared in a more generic framework of guiding assumptions. Whichever theory that explains the historical events best should be considered as the superior theory. This idea was later criticized as being too ambitious (Radder, 1997).

Kuhn's theory of scientific revolutions describes how science advances through a path of normal science, crisis, revolution, and new normal science (Kuhn, 1962, 1970). A revolution is a Gestalt shift of world views. Criticisms of the theory argue that scientific change often takes place through a lengthy process instead of a swift change as the paradigm-shift model suggested. Although Kuhn made his cases with groundbreaking grand revolutions such as the Copernican model of the solar system and Einstein's relativity theory, his theory is generally applicable to revolutions of much smaller scale. Van Raan indicated that cocitation clusters may be characterized by power laws, suggesting a wide spectrum of revolutions.

Henry Small studied the example of atomic physics in the early twentieth century. He noticed that until Niels Bohr's 1913 model for the hydrogen atom using a quantum hypothesis, there was no direct connection between experimental evidence on the spectrum for atomic hydrogen and evidence for hydrogen's nuclear structure in the relevant literature. Similarly, the Müller-and-Bednorz discovery of superconductivity was made by bridging a gap between the knowledge of superconductivity and compounds that were never thought to have anything to do with superconductivity (Holton, Chang, & Jurkowitz, 1996; Small, 2000).

Burt's structural hole theory is well known in the context of social networks (Burt, 1992, 2004, 2005). A structural hole in a network is a sparse subnetwork. Members of the sparse subnetworks are not connected at all or connected via a small number of paths. Spanning a structural hole means adding new paths to the subnetwork so that previously disparate nodes become connected or previously underconnected nodes become better connected. Burt argued that since information flow has more constraints due to the connectivity across a structural hole, those in a brokerage position between different constituencies of information tend to have advantages over others in the network. Brokerage positions yield a competitive edge. We adopt the structural hole theory for networks of cited references and propose that adding new links over a structural hole is a mechanism for making new discoveries. In later chapters, we will see that similar ideas have been proposed to characterize the creation of radical patents through a recombinant process.

A recurring theme is that profound scientific changes often rise from the work of a consistent range of boundary spanning, bridge building, or brokerage activities. The greatest philosophers are often the ones who were at the center of battles with competing schools of thought. Creative artists are found to be the ones who stay in touch with artists from different circles (Guiffre, 1999). Creative scientists are more likely to maintain connections with peer scientists who belong to diverse and disparate groups (Heinze & Bauer, 2007). Scientists make extra efforts to maintain contacts with scientists in different fields (Crane, 1972). The greatest discoveries come from no-man's land (Wiener, 1948).

Our central premise is therefore that bridging gaps in a knowledge space is a valuable and viable mechanism for understanding and achieving transformative scientific discoveries.

4.2 CITESPACE

CiteSpace[1] is an interactive computer application developed to facilitate the study of structural and temporal trends and patterns in scientific literature. Its major functions are designed to visualize and analyze the dynamics of a scientific knowledge domain (Chen, 2004, 2006; Chen, Ibekwe-SanJuan, & Hou, 2010). The CiteSpace system provides a readily accessible tool for a wide variety of users to interact with scientific literature through interactive visualizations of scientific knowledge. It supports the entire workflow of generating a systematic review of a knowledge domain directly from the relevant literature.

---

[1] http://cluster.ischool.drexel.edu/~cchen/citespace

Derek Price, a pioneer of scientometrics—the quantitative study of science—noticed that two major streams of publications contribute to the growth of scientific literature at different rates. Original research papers appear in the scientific literature at a much faster rate than systematic review and survey papers. According to Price's observation, after the publication of about 50 original research articles in a field of study, a review article is likely to appear. Review articles summarize and consolidate the contributions of these original research articles and their significance.

A comprehensive review article is in general valuable to the development of a research field, but for a particular individual researcher, the most desirable reviews may be not accessible. Review articles require a substantial degree of domain expertise and an insightful understanding of the past, the present, and the future of the field. Domain experts who have the knowledge and time to produce a review of the field may decide that it is not yet the time to write a review article. Besides, all domain experts have their own preferences in terms of where they prefer to direct their attention across an array of subtopics. After all, even if they write it, they may not approach it from a point of view that you would like to see.

CiteSpace is to provide an interactive visual analytics system to a wide variety of users so that each individual can generate an overview map of a knowledge domain. The overview map serves the same purpose as a systematic review of the literature. The user will see from the overview map and associated details how a knowledge domain is self-organized, namely, how many components in the system as a whole, how often a previously published work is cited and followed by subsequent research, how these components are connected, and which publications play a critical role in the evolution of the domain.

The other motivation is to provide an analytic tool for the study of the structure of scientific revolutions as the way conceptualized by Thomas Kuhn. The tool facilitates scholars to analyze the evolution of a scientific knowledge domain. The results are more comparable than that of traditional historical analyses. Finally, it is the foresight. By supporting predictive analytics of scientific literature, the system detects early signs that may help us identify promising paths of research.

4.2.1 Design Rationale

CiteSpace has been evolving since its initial publication in 2004 in the *Proceedings of the National Academy of the United States of America*. The current architecture of CiteSpace is shown in Figure 4.5. The primary source of input is the scientific literature, especially bibliographic records with cited references retrieved from the SCI expanded through the Web of Science. The user can select various control parameters to generate interactive visualizations at global and local levels (more details shortly).

CiteSpace supports interactive visualizations of global and local maps of scientific knowledge. The global map consists of two base maps of scientific journals. Each map depicts over 10,000 journals. These journals are organized based on their citation patterns. In the citing map, journals are organized based on their citing patterns. In the cited map, journals are organized based on their cited patterns. Users can explore multiple datasets of their interest in the form of interactive overlays. Each interactive

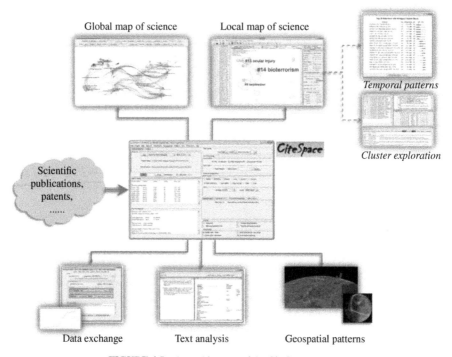

FIGURE 4.5   An architecture of the CiteSpace system.

overlay is a layer of visualization superimposed on top of the two base maps such that users can trace paths of citations from the sources of references to the destinations. Users can also study the movement of the publication activities as a whole over the maps. Since the geography of the base maps reveals the disciplinary boundaries, such movements can help users to understand the dynamics of publication portfolios for individuals, organizations, and other types of entities defined by the datasets. For example, a publication portfolio consists of all the publications with authors affiliated with the university. An analyst can compare multiple portfolios and identify specifically how these universities differ in terms of where they publish and where they draw their inspirations. We will describe the design and use of global overlay maps in later chapters after we establish relevant conceptual frameworks.

The local map in CiteSpace provides user-driven interactive visualizations of scientific literature based on a single dataset. An input dataset is defined by the user, who is responsible for the scope of the dataset. The scope of a valid dataset ranges from a focused concentration of publications relevant to mad cow diseases to a comprehensive collection of publications that represent the research activity of a discipline or a field such as astronomy or medicine. The structure and dynamics of scientific literature are visualized for interactive exploration and reporting (upper right in Figure 4.5). CiteSpace also supports text analysis and geospatial visualizations (the bottom of Figure 4.5). In addition to scientific literature, the visual analytics procedure can be applied to other types of data such as U.S. patents, U.S. Supreme Court Opinions, and the National Science Foundation (NSF) awards.

CiteSpace supports users to perform a series of steps to analyze structural and temporal patterns in scientific literature. Users may address a variety of analytic questions with the tool. Here are some examples:

- What are the landmark publications in the chosen knowledge domain?
- What are the major areas of the domain?
- How are these major areas connected?
- Which articles are critical to the structure of the domain as a whole?
- Which areas are currently attracting the most attention?

We use cocitations of references as the basic mechanism to pull relevant articles from the literature and construct a global structure from local details. Each individual scientist or domain expert may influence the structure of the scientific literature with the way they cited references in the literature. Their choices of citation may reinforce or alter the current connectivity among the available references. All the publications in the literature form a complex adaptive system. Local variations may lead to global, system-level changes.

4.2.2 Basic Procedure

The procedure of generating local maps in CiteSpace is shown in Figure 4.6. CiteSpace divides the entire window of analysis into a sequence of consecutive time intervals, called time slices. Citation behaviors observed within each time slice are used to construct a network model. Networks over adjacent time slices are merged to form a network over a longer period of time.

Synthesized networks can be divided into clusters of cocited references. Each cluster contains a set of references. The formation of a cluster is resulted from the citation behaviors of a group of scientists who are concerned with the same set of research problems. A cluster therefore can be seen as the footprint of an invisible college. As the invisible college changes its research focus, its footprints will move on the landscape of scientific knowledge. The cluster will evolve accordingly. For example, it may continue to grow in size, branch out to several smaller clusters, or join other clusters. It may even be phased out as the invisible college drift away from an old line of research altogether.

CiteSpace provides three algorithms to label a cluster, namely, the traditional *tf* by *idf*, log-likelihood ratio (LLR), and mutual information (MI). Label terms are selected from titles, keywords, or abstracts of articles that specifically cite members of the cluster. If the members of a cluster represent the footprints of an invisible college or a paradigm, the labels reflect what the invisible college and the paradigm are currently concerned with, which may or may not be consistent with the direction of the cluster. These clusters represent the intellectual base of a paradigm, whereas the citing articles associated with a cluster represent the research fronts. It is possible that the same intellectual base may sustain more than one research front.

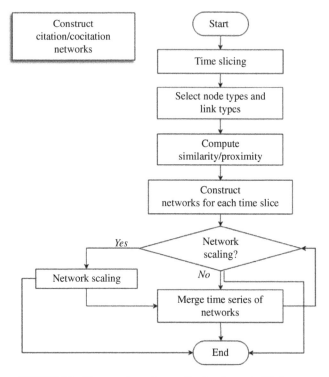

FIGURE 4.6 The procedure of generating local maps in CiteSpace.

Time Slicing

Time slicing divides the entire time interval into equal-length segments called time slices. The duration of each segment can be as short as one year or as long as tens and even hundreds of years. If appropriate data are available, it is possible to slice it thinner to make monthly or weekly segments. Currently, sliced segments are mutually exclusive, although overlapping segments could be an interesting alternative worth exploring.

Sampling

Citation analysis and cocitation analysis typically sample the most highly cited work—the cream of the crop. In order to construct a network in CiteSpace, users may set their own criteria for node selection and link selection. Alternatively, they can use the default setting provided by CiteSpace. The simplest way to select nodes is the *Top-N* method, in which the *N* of the most cited articles within the timeframe of each slice will be included in the final network. Similarly, the *Top-N%* method will include the *N%* of the most cited references within each slice. CiteSpace also allows the user to choose three sets of threshold values and interpolates these values across all the slices. Each set of threshold values is as follows: a citation count (c), a cocitation count (cc), and a cosine coefficient of cocitation similarity (ccv). In *CiteSpace*, the user needs to select desired thresholds in the beginning, the middle, and the ending slices. CiteSpace automatically assigns interpolated thresholds to the remaining slices.

Research has shown that citation counts often follow a power law distribution. The vast majority of published articles are never cited. On the other hand, a small number of articles dominate a lion's share of citations. Many factors may influence the frequency and distribution of citations to published articles. A highly cited article is highly visible. Its visibility is likely to attract more citations. As far as intellectual turning points are concerned, we are particularly interested in articles that have rapidly growing citations. In the following superstring example, we use a simple model to normalize the citations of an article within each time slice by the logarithm of its publication age—the number of years elapsed since its publication year. The rationale is to highlight articles that increased most in the early years of their publication.

Modeling

By default, cocitation counts are calculated within each time-sliced segment. Cocitation counts are normalized as cosine coefficients, provided $c(i) \neq 0$ and $c(j) \neq 0$:

$$cc_{cosine}(i, j) = \frac{cc(i, j)}{\sqrt{c(i) * c(j)}} \tag{4.3}$$

where $cc(i, j)$ is the cocitation count between documents i and j, and $c(i)$ and $c(j)$ are their citation counts, respectively. The user can specify a selection threshold for cocitation coefficients; the default value is 0.15. Alternative measures of cocitation strengths are available in the information science literature, such as Dice and Jaccard coefficients.

Pruning

An effective pruning can reduce link crossings and improve the clarity of the resultant network visualization. CiteSpace supports two common network pruning algorithms, namely, Pathfinder and MST. The user can select to prune individual networks only, or the merged network only, or prune both. Pruning increases the complexity of the visualization process. In the following section, visualizations with local pruning and global pruning are presented.

Here we concentrate on Pathfinder-based pruning. To prune individual networks with Pathfinder, the parameters q and r were set to $N_k - 1$ and ∞, respectively, to ensure the most extensive pruning effect, where N_k was the size of the network in the kth time slice. For the merged network, the q parameter was $(\Sigma N_k) - 1$, for $k = 1, 2, \ldots$

Merging

The sequence of time-sliced networks is merged into a synthesized network, which contains the set union of all nodes ever appear in any of the individual networks. Links from individual networks are merged based on either the earliest establishment rule or the latest reinforcement rule. The earliest establishment rule selects the link that has the earliest time stamp and drops subsequent links connecting the same pair of nodes, whereas the latest reinforcement rule retains the link that has the latest time stamp and eliminates earlier links.

By default, the earliest establishment rule applies. The rationale is to support the detection of the earliest moment when a connection was made in the literature. More precisely, such links mark the first time a connection becomes strong enough with respect to the chosen thresholds.

Visual Encoding

The layout of each network, either individual time-sliced networks or the merged one, is produced using Kamada and Kawai's algorithm (Kamada & Kawai, 1989). The size of a node is proportional to the normalized citation counts in the latest time interval. Landmark nodes can be identified by their large disks. The label size of each node is proportional to citations of the article, thus larger nodes also have larger-sized labels. The user can enlarge font sizes at will. Both the width and the length of a link are proportional to the corresponding cocitation coefficient. The color of a link indicates the earliest appearance time of the link with reference to chosen thresholds. Visually salient nodes such as landmarks, hubs, and pivots are easy to detect by visual inspection and enhanced computationally by algorithms.

CiteSpace provides three types of visualizations, namely, a cluster view, a timeline view, and a timezone view. In the timeline view, clusters are displayed horizontally alone timelines. The label of each cluster is shown at the end of the cluster's timeline. Cited references or authors are depicted as circles filled with citation rings. The color of each ring corresponds to the time slice in which citations were made. The thickness of a ring is proportional to the amount of citations received in that time slice. Thus, a large-sized circle denotes a highly cited unit, that is, reference or author.

The colors of red and purple are used to highlight special attributes of a node. A red ring indicates that a citation burst is detected in the corresponding time slice. A purple ring is added to a node if its betweenness centrality is greater than 0.1. The thickness of the ring is proportional to its centrality value.

A line connecting two items in the visualization represents a cocitation link. The thickness of a line is proportional to the strength of cocitation. The color of a line represents the time slice in which the cocitation was made for the first time. A useful byproduct of spectral clustering is that tightly coupled clusters tend to be placed next to each other and visually form a supercluster.

4.2.3 Advanced Cocitation Analysis

The traditional approach to the study of cocitation networks provides one of the many possible interpretations of the impact of scientific publications, but a direct link may still be missing. Using a multiple-perspective approach, we intend to uncover such links that have been overlooked in traditional cocitation studies. In particular, direct connections are established by the support of three analytic tasks, namely, clustering, labeling, and sentence selection.

The multiple-perspective approach extends and enhances traditional cocitation methods in two ways: (1) by integrating structural and content analysis components sequentially into the new procedure and (2) by facilitating analytic tasks and interpretation with automatic cluster labeling and summarization functions. The key

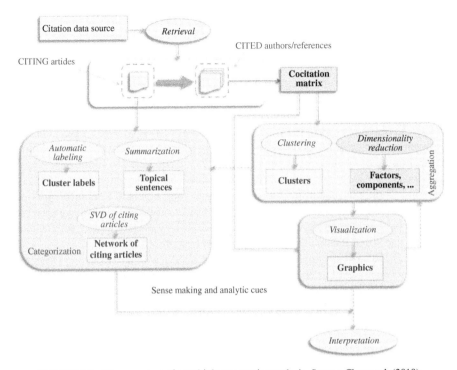

FIGURE 4.7   The procedure of a multiple-perspective analysis. Source: Chen et al. (2010).

components of the new procedure are shown in Figure 4.7, including clustering, automatic labeling, summarization, and latent semantic models of the citing space (Deerwester, Dumais, Landauer, Furnas, & Harshman, 1990).

The primary goal of cocitation analysis is to identify the intellectual structure of a scientific knowledge domain in terms of the groupings formed by accumulated cocitation trails in scientific literature. The traditional procedure of cocitation analysis for both document cocitation analysis (DCA) and author cocitation analysis (ACA) consists of the following steps:

1. Retrieve citation data from sources such as the *SCI*, *SSCI*, *Scopus*, and *Google Scholar*.
2. Construct a matrix of cocited references (DCA) or authors (ACA).
3. Represent the cocitation matrix as a node-and-link graph or as an MDS configuration with possible link pruning using Pathfinder network scaling or MST algorithms.
4. Identify specialties in terms of cocitation clusters, multivariate factors, principle components, or dimensions of a latent semantic space using a variety of algorithms for clustering, community finding, factor analysis, PCA, or latent semantic indexing.
5. Interpret the nature of cocitation clusters.

The weakest link of the procedure is the interpretation step. It is time consuming and cognitively demanding, requiring a substantial level of domain knowledge and synthesizing skills. In addition, since much of attention routinely focuses on interpreting the impact of cocitation clusters per se without explicit supporting information, such interpretations may overlook the actual formation of such cocitation clusters. For example, Galea et al.'s article on the prevalence of PTSD after the September 11, 2001, terrorist attacks discovered an unprecedented observation of the prevalence of PTSD in adults who were not directly involved with trauma. Their article cited 32 references, but none of them reported the new link they discovered. A traditional cocitation study would focus on interpreting various clusters formed by the 32 previous references and other relevant references. Since the information about the new link is only available in Galea et al.'s article and nowhere else in the previous literature, such interpretations would have no way to bridge the gap between the past and the present. Therefore, to resolve the issue, it is necessary to take into account articles that shape the cocitation clusters in the first place.

To help the analyst estimate the quality of an overview of the literature, CiteSpace provides system-level measures of the homogeneity and consistency of all the component clusters as an indicator of both the overall quality and the quality of each individual cluster. Silhouette scores are computed for this purpose. The silhouette value of a cluster ranges from −1 to 1. The value of a homogeneous cluster is close to 1, which means the quality of the grouping is high. Since each network may be divided into clusters in many ways, the most plausible interpretation should be based on the decomposition that strikes a balance between the modularity of the division and the local homogeneity of each divided component.

Citation behaviors are complex and diverse (Cronin, 1981). Synthesizing the nature of a cocitation cluster manually is impractical due to the demanding cognitive burden on the analyst. The lack of algorithmic support for these tasks is likely to force analysts to resort to their own domain knowledge and their experience, which in turn would make their interpretation more subjective than they intended. Such ambiguity and subjectivity may hinder subsequent evaluation and scholarly communication of research findings.

Metrics

Structural metrics such as the betweenness centrality of a node, the modularity, and the average silhouette of a network's decomposition provide valuable indicators for us to identify important information.

The betweenness centrality of a node in a network measures the likelihood that the node is part of a path in the network (Brandes, 2001; Freeman, 1977). Since a path connects a source node s and a target node t in the network, for a node n to be part of the path, the node is connected to both of the other two nodes. Two scenarios are particularly interesting. One scenario is that s belongs to a large and dense subnetwork S, and similarly t belongs to a large subnetwork of T, but S and T are not connected except via n. In other words, n is the only way S and T are connected. The betweenness centrality of n must be high and it is in-between S and T. In another scenario, the betweenness centrality of n is high if it is the center of the entire network.

Since the scientific community is likely to be aware of the central piece in its literature, we are more interested in drawing people's attention to nodes described in the first scenario. In the first scenario, a high-betweenness centrality node is a good candidate for a potentially revolutionary scientific publication (Chen, 2005). Power centrality introduced by Bonacich (1987) is another option to capture the concept of centrality (Kiss & Bichler, 2008).

The modularity Q measures the extent to which a network can be divided into independent blocks, that is, modules (Newman, 2006; Shibata, Kajikawa, Taked, & Matsushima, 2008). The modularity score ranges from 0 to 1. A low modularity suggests a network that cannot be reduced to clusters with clear-cut divisions, whereas a high modularity may imply a well-connected network. Networks with modularity scores of 1 or very close to 1 may become trivial when individual components degenerate to singleton nodes that are simply isolated from one another.

The silhouette metric (Rousseeuw, 1987) is useful in estimating the uncertainty involved in identifying the nature of a cluster. The silhouette value of a cluster, ranging from −1 to 1, indicates the uncertainty that one needs to take into account when interpreting the nature of the cluster. The value of 1 represents a perfect separation from other clusters. In this study, we expect that cluster labeling or other aggregation tasks will become more straightforward for clusters with the silhouette value in the range 0.7–0.9 or higher.

Burst detection determines whether a given frequency function has statistically significant fluctuations during a short time interval within the overall time period. It is valuable for citation analysts to detect when and whether the citation count of a particular reference has surged. For example, after the September 11 terrorist attacks, citations to earlier studies of Oklahoma City Bombing increased abruptly (Chen, 2006). It can be also used to detect whether a particular connection has been significantly strengthened within a short period of time (Kumar, Novak, Raghavan, & Tomkins, 2003). We adopt the burst detection algorithm introduced in Kleinberg (2002).

Sigma (Σ) is introduced in Chen, Zhang, and Vogeley (2009) as a measure of scientific novelty. It identifies scientific publications that are likely to represent novel ideas according to two criteria of transformative discovery. As demonstrated in case studies (Chen, Chen et al., 2009), Nobel Prize and other award-winning research tends to have highest values of this measure. CiteSpace currently uses (centrality + 1)^{burstness} as the Σ value so that the brokerage mechanism plays more prominent role than the rate of recognition by peers.

Clustering

We adopt a hard clustering approach such that a cocitation network is partitioned to a number of nonoverlapping clusters. It is more efficient to use nonoverlapping clusters than overlapping ones to differentiate the nature of different cocitation clusters, although it is conceivable to derive a soft clustering version of this particular component. Resultant clusters are subsequently labeled and summarized.

Cocitation similarities between items i and j are measured in terms of cosine coefficients. If A is the set of papers that cites i and B is the set of papers that cite j, then

$w_{ij} = |A \cap B| / \sqrt{|A| \times |B|}$, where $|A|$ and $|B|$ are the citation counts of i and j, respectively, and $|A \cap B|$ is the cocitation count, that is, the number of times they are cited together. Alternative similarity measures are also available. For example, Small (1973) used $w_{ij} = |A \cap B| / |A \cup B|$, which is known as the Jaccard index (Jaccard, 1901).

A good partition of a network would group strongly connected nodes together and assign loosely connected ones to different clusters. This idea can be formulated as an optimization problem in terms of a cut function defined over a partition of a network. Technical details are given in relevant literature (Luxburg, 2006; Nikkila et al., 2002; Shi & Malik, 2000). A *partition* of a network G is defined by a set of subgraphs $\{G_k\}$ such that $G = \bigcup_{k=1}^{K} G_k$ and $G_i \cap G_j = \emptyset$, for all $i \neq j$. Given subgraphs A and B, a *cut function* is defined as follows: cut $(A, B) = \sum_{i \in A, j \in B} w_{ij}$, where w_{ij} is the cosine coefficient mentioned earlier. The criterion that items in the same cluster should have strong connections can be optimized by maximizing $\sum_{k=1}^{K} \text{cut}(G_k, G_k)$. The criterion that items between different clusters should be only weakly connected can be optimized by minimizing $\sum_{k=1}^{K} \text{cut}(G_k, G - G_k)$. In this study, the cut function is normalized by $\sum_{k=1}^{K} (\text{cut}(G_k, G - G_k) / \text{vol}(G_k))$ to achieve more balanced partitions, where $\text{vol}(G_k)$ is the sum of the weights of links in G_k, that is, $\text{vol}(G_k) = \sum_{i \in G_k} \sum_{j} w_{ij}$ (Shi & Malik, 2000).

Spectral clustering is an efficient and generic clustering method (Luxburg, 2006; Nikkila et al., 2002; Shi & Malik, 2000). It has roots in spectral graph theory. Spectral clustering algorithms identify clusters based on eigenvectors of Laplacian matrices derived from the original network. Spectral clustering has several desirable features compared to traditional algorithms such as k-means and single linkage (Luxburg, 2006):

1. It is more flexible and robust because it does not make any assumptions on the forms of the clusters.
2. It makes use of standard linear algebra methods to solve clustering problems.
3. It is often more efficient than traditional clustering algorithms.

The multiple-perspective method utilizes the same spectral clustering algorithm for both ACA and DCA studies. Despite its limitations (Luxburg, Bousquet, & Belkin, 2009), spectral clustering provides clearly defined information for subsequent automatic labeling and summarization to work with. In this study, instead of letting the analyst to specify how many clusters there should be, the number of clusters is uniformly determined by the spectral clustering algorithm based on the optimal cut described earlier.

Automatic Cluster Labeling
Candidates of cluster labels are selected from noun phrases and index terms of citing articles of each cluster. These terms are ranked by three different algorithms.

In particular, noun phrases are extracted from titles and abstracts of citing articles. The three term-ranking algorithms are $tf \times idf$ (Salton, Yang, & Wong, 1975), *LLR* tests (Dunning, 1993), and *MI*. Labels selected by $tf \times idf$ weighting tend to represent the most salient aspect of a cluster, whereas those chosen by LLR tests and MI tend to reflect a unique aspect of a cluster.

Garfield (1979) has discussed various challenges of computationally selecting the most meaningful terms from scientific publications for subject indexing. Indeed, the notion of citation indexing was originally proposed as an alternative strategy to deal with some of the challenges. White (2007a, 2007b) offers a new way to capture the relevance of a communication in terms of the widely known $tf \times idf$ formula.

A good text summary should have a sufficient and balanced coverage with minimal redundant information (Sparck-Jones, 1999). Teufel and Moens (2002) proposed an intriguing strategy for summarizing scientific articles based on the rhetorical status of statements in an article. Their strategy specifically focuses on identifying the new contribution of a source article and its connections to earlier work. Automatic summarization techniques have been applied to areas such as identifying drug interventions from MEDLINE (Fiszman, Demner-Fushman, Kilicoglu, & Rindflesch, 2009).

Modeling Structural Variations

An important question is how newly received information is likely to change the existing structure of a network. Every network has its own context or environment. If changes take place in its environment, it is often necessary to ask whether these changes in the environment have any impact on the structure of the network. In light of new information or new evidence, it may become necessary to update the structure of the existing network, especially when it may lead to a significantly different structure.

To detect the novelty of a new scientific publication with respect to the current structure of knowledge, we focus its effect on structural variation resulted from its boundary spanning activities. Consider the following examples. Which one do you take for granted, and which one would surprise you?

1. *soccer ~ beer*
2. *soccer ~ octopus*

The connection between soccer and beer is well established—every four years football fans all over the world watch FIFA's world cup matches and many of them would go and watch the games in bars, pubs, or anywhere in the proximity of beers. In contrast, the word *octopus* had little to do with the sport of soccer—until the 2010 FIFA World Cup in South Africa. A particular octopus was under the spotlight before increasingly important games because it mysteriously managed to predict the winners of many games so accurately. The connection between soccer and octopus was novel because prior to the 2010 World Cup, most people would not think an octopus would have any association with soccer.

There are numerous examples of how new theories revolutionized the contemporary knowledge structure. The 2005 Nobel Prize in medicine, for example, was awarded to the discovery of *Helicobacter pylori*, a bacterium that was previously

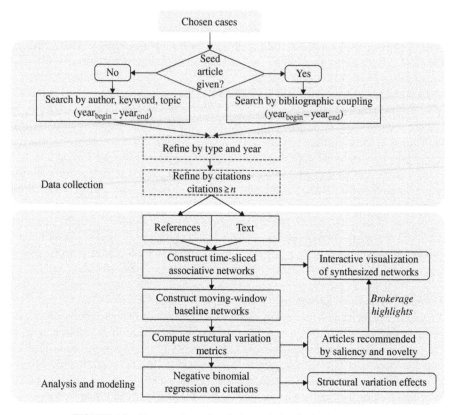

FIGURE 4.8   The procedure of predictive analysis of scientific literature.

believed not to be able to survive in human's gastric system (Chen, Chen et al., 2009). In literature-based discovery, Swanson discovered previously unnoticed linkage between fish oil and Raynaud's syndrome (Swanson, 1986). Therefore, a fundamental question concerning the value of new information is: to what extent is the newly available information likely to change what we believe? The procedure of conducting a predictive analysis in terms of structural variations is shown in Figure 4.8.

4.2.4   Toward a Tightly Connected Community

In order to improve our understanding of an interactive visual analytic process in the context of a study of scientific literature, we conceptualize the engagement of a user with a visual analytic system such as CiteSpace as a process in a state transition space. For example, a user may perform a clustering analysis and then explore each cluster. The sequence of corresponding events represents a series of transitions from one state of the analysis to another. An individual user's behavior is therefore recorded in this way for subsequent analysis. A complete user session can be represented by a matrix. The moves made by the user are represented by the strengths of links between corresponding states (vertices). The behavioral patterns of a user community can be identified by aggregating individual matrices. Users with similar transition patterns can be identified. The differences between experienced users and novices can be highlighted. These additional sources of information provide users a new learning opportunity to reflect on their own navigation choices through the user–system interaction.

Interactive Events

A trail of a typical series of interactive events is illustrated in Figure 4.9. The user started by configuring the time interval for analysis (1998–2012) and moved to the node selection by selecting the top 20 highly cited references in each year. The user then selected DCA for the type of analysis to perform and applied Pathfinder as the link reduction before invoking the process at the GO button. Once the visualization was available, the user selected the option to view. The events shown in the boxes are collected from the session. Prior to each session, the consent is obtained explicitly from the user for recording such events for scholarly analysis. Figure 4.10 shows another segment of a sequence of interactive events as the user explored control functions in CiteSpace. The thick arrow indicates that the user repeatedly clicked through multiple time slices to reveal how the network evolves year by year.

The results of a pilot test suggested that users' interactive events provide valuable information on whether a user is learning about a specific group of analytic functions or exploring the capability of the system in general. Table 4.2 reveals that researchers from at least three different cities, Guiyang, Changsha, and Shanghai, are simultaneously engaged in the study of sports and closely related topics such as the Olympic Games. Geographically, these cities are not in close proximity to one another. The distance between Guiyang and Changsha is 520 miles. The distance between Shanghai and Changsha is 683 miles. The distance between Guiyang and Shanghai is 1153 miles. The apparent common interest in analyzing the literature of sports is a potential common ground to forge a virtual community so that researchers from different cities can share and exchange their ideas and experiences with the study of the same literature.

A potentially useful indicator that may differentiate an experienced user from a novice user is the average size of the networks they have been analyzing. A novice user is usually reluctant to take on a large network before they have a good idea of how the system would behave in general. Since it is possible for the system to trace the change of user behavior over time, it would be informative for the user to learn about various behavioral changes.

CiteSpace is continuously evolving. Logs of users' analytic events inform us how soon users are likely to switch to a new version of the software, how long some other users would keep using an old version, and which cities and countries these users are located. This type of questions will improve our understanding of how information technology is diffused in an international user community.

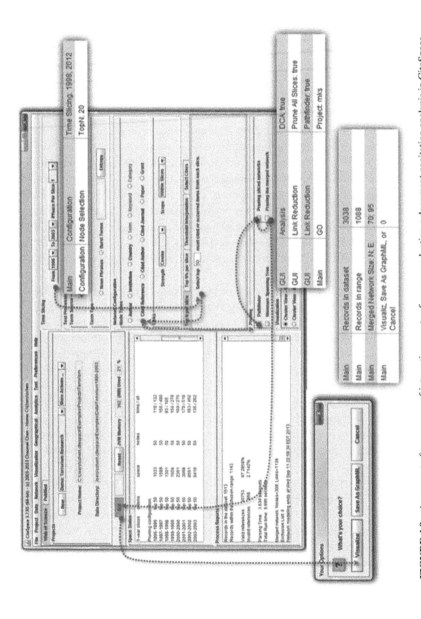

FIGURE 4.9 A segment of a sequence of interactive events of a user conducting a document cocitation analysis in CiteSpace.

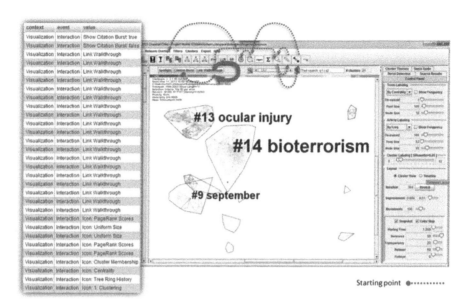

FIGURE 4.10   A segment of a sequence of interactive events as a user exploring various control functions in CiteSpace.

Table 4.2   Users from different cities may study the same topics

Freq	IP (partial)	The largest cluster	Location
3	111.85 ___.___.	Sport	Guiyang, Guizhou, China
1	111.85 ___.___.	Panhellenic game; aggressive sports; catharsis;	Guiyang, Guizhou, China
1	58.16 ___.___.	Olympic games; American; China	Guiyang, Guizhou, China
2	118.239 ___.___.	Sport	Changsha, Hunan, China
1	118.239 ___.___.	Olympics; amateurism; Celtic nationalism; British	Changsha, Hunan, China
2	118.239 ___.___.	Olympics; event-planning; amateurism; Celtic nationalism	Changsha, Hunan, China
1	118.239 ___.___.	Panhellenic game; aggressive sports; catharsis	Changsha, Hunan, China
1	118.239 ___.___.	Sport	Changsha, Hunan, China
4	118.239 ___.___.	Britain; sport	Changsha, Hunan, China
1	114.91 ___.___.	Sport	Shanghai, China

Build a Tightly Connected Community

The widely spread user population has the potential to evolve and transform itself to a tightly connected community. The members of the community will be able to engage a wider variety of learning and collaborative activities with other users. These peer learners and potential collaborators may have no other way to get to know each other because what they have in common may be only in terms of how they interact with

CiteSpace or the topic they choose to analyze with CiteSpace. A tightly connected community would provide users additional opportunities to learn and collaborate.

To foster a sense of community, an awareness of concurrent users would be provided as a new feature to the community edition of CiteSpace. A number of proximity measures will be generated for each user so that a user would be aware if another user is out there using CiteSpace in a similar area of functionality, analyzing publications on a similar topic, or being an expert user in the same city. Perhaps the most valuable addition would be the possibility for experienced CiteSpace users to provide instance assistance to concurrent users. The community platform would be an ideal vehicle for online tutorials and workshops.

A generalized notion of proximity is defined to inform a user about concurrent or recent users who may be a potentially helpful source of information for learning or collaboration. A *geographic proximity* is defined as the geographic distance between the locations of two users. To make it easy to understand, a geographic proximity can simply indicate whether the two users are in the same city, the same country, or the same continent instead of using a numeric measure. An *analytic proximity* is defined as the difference between two users' state transition matrices, for example, in terms of the associated relative entropy, or the Kullback–Leibler divergence. A *topical proximity* is defined as the topical similarity between the datasets that two users are analyzing or recently analyzed. For example, several users from different cities in China are analyzing scientific literature on sports because the largest clusters in their analyses are labeled by terms relevant to sports. If they become aware of the common interest, they may express their interest in sharing their experience, exchanging findings, and exploring opportunities for collaboration.

4.3   EXAMPLES

We have studies a series of cases in several disciplines using CiteSpace, including mass extinctions, terrorism research (Chen, 2006), SDSS in astronomy (Chen et al., 2009), information science (Chen et al., 2010), and regenerative medicine (Chen, Hu, Liu, & Tseng, 2012). Researchers from different countries have used CiteSpace in their own research. Here we outline some of the examples of our own studies.

4.3.1   Terrorism Research

The terrorism research case is one of the earliest cases we developed with the initial version of CiteSpace. We have subsequently revisited the case a number of times as a standard reference to test various new features. The case is included in the standard distribution of CiteSpace. Since the case has been analyzed in detail, it provides a suitable starting point to explore the process of visual analytics of scientific literature.

Figure 4.11 shows a visualization of the terrorist research case. The overall network is dominated by three clusters of densely cocited references. Each cluster represents a paradigm. The one at the bottom was formed after the terrorist attacks on September 11, 2001. The one on the left was formed much earlier, which is primarily

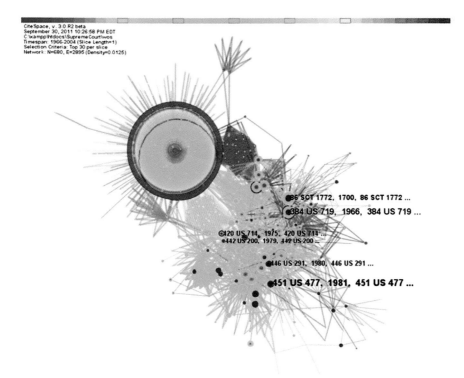

CiteSpace, v. 3.0 R2 beta
September 30, 2011 10:26:58 PM EDT
C:\xampp\htdocs\SupremeCourt\wos
Timespan: 1966-2004 (Slice Length=1)
Selection Criteria: Top 30 per slice
Network: N=680, E=2895 (Density=0.0125)

86 SCT 1772, 1700, 86 SCT 1772 ...

384 US 719, 1966, 384 US 719 ...

420 US 714, 1975, 420 US 714 ...
442 US 200, 1979, 442 US 200 ...

446 US 291, 1980, 446 US 291 ...

451 US 477, 1981, 451 US 477 ...

FIGURE 3.13   A visualization of a cocitation network of cases cited with the case *Miranda v. Arizona*, 384 U.S. 436 (1966). The size of a circle is proportional to the citations received by the case. Red circles represent cases with citation bursts.

FIGURE 4.2   An animated visualization of a network-based landscape view of the literature on mad cow disease.

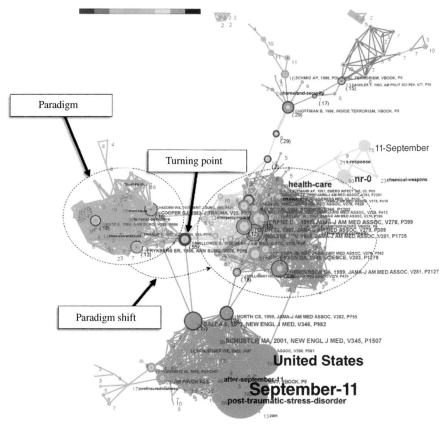

FIGURE 4.11 Major areas in terrorism research. Source: Chen (2006).

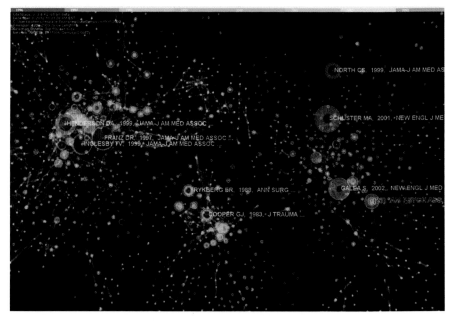

FIGURE 4.13 A visualization of a 3638-node network generated by sampling top 500 most cited articles each year between 1996 and 2003.

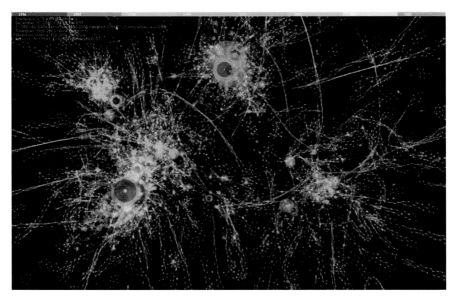

FIGURE 4.14 A network of 12,691 cocited references generated by sampling top 2000 most cited articles per year.

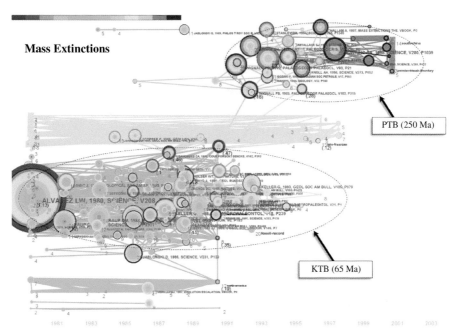

FIGURE 4.15 Trends in mass extinctions research. Source: Chen (2006).

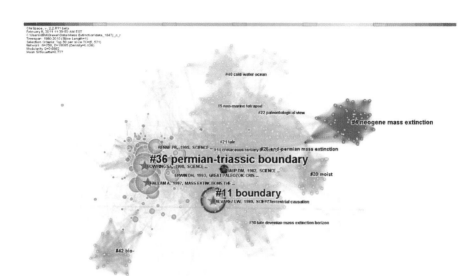

CiteSpace, v. 2.2.R11 beta
February 8, 2011 11:39:50 AM EST
C:\Users\BliAGraeat\Data\Mass Extinction\data_1847)_n_r
Timespan: 1980-2010 (Slice Length=1)
Selection Criteria: Top 50 per slice TC=(8_571)
Network: N=259, E=10685 (Density=0.036)
Modularity Q=0.8862
Mean Silhouette=0.777

#40 cold water ocean

#5 neo-marine tetrapod

#22 paleontological view

#4 neogene mass extinction

#21 late

RENNE PR, 1995, SCIENCE ... #11 cretaceous tertiary #26 and-permian mass extinction

#36 permian-triassic boundary

BOWRING SA, 1998, SCIENCE ...

RAUP DM, 1982, SCIENCE ...

ERWIN DH, 1993, GREAT PALEOZOIC CRIS ...

#39 moist

HALLAM A, 1997, MASS EXTINCTIONS THE ...

#11 boundary

ALVAREZ LW, 1980, SCIE#7Cterrestrial causation

#30 late devonian mass extinction horizon

#42 bio-

FIGURE 4.17   A visualization of mass extinction in a follow-up study in 2011.

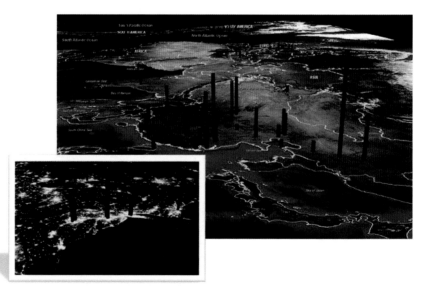

FIGURE 4.19   The distribution of CiteSpace's users. The height of a bar represents the average size of networks analyzed by users at a city.

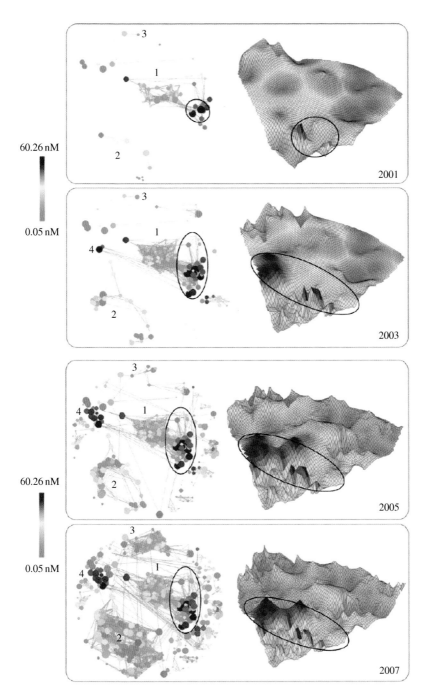

FIGURE 5.3 The evolution of the AA2 dataset is shown in four snapshots representing the structure–activity landscape in 2001, 2003, 2005, and 2007, respectively. Source: Reprinted with permission from Figure 3a in Iyer, Hu, and Bajorath (2011). Copyright (2011) American Chemical Society.

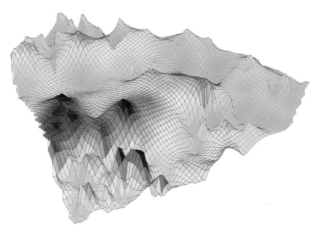

FIGURE 5.4 A close-up view of three-dimensional SAR landscape of the AA2 dataset in 2005. Source: Reprinted with permission from Figure S1 2005 in Iyer et al. (2011). Copyright (2011) American Chemical Society. http://pubs.acs.org/doi/suppl/10.1021/ci100505m/suppl_file/ci100505m_si_001.pdf.

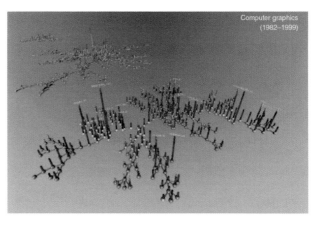

FIGURE 5.8 A rising landscape of research about mad cow disease (green), CJD in human (blue), and the fundamental research on prions (red).

FIGURE 5.9 Thematic landscapes of computer graphics (1982–1999).

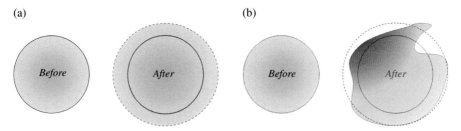

FIGURE 6.3 Incremental changes (a) and transformative changes (b) caused by a perturbation to the system.

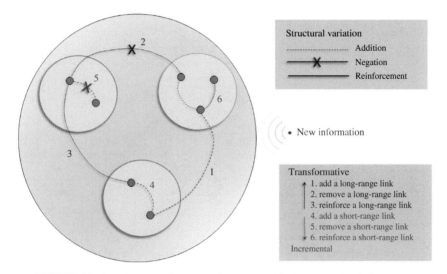

FIGURE 6.4 Boundary-spanning mechanisms modeled in the structural variation theory.

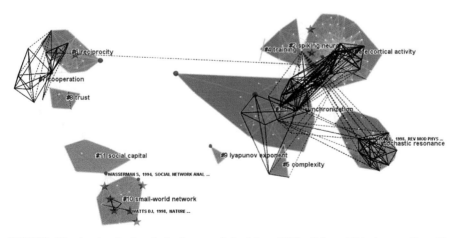

FIGURE 6.7 A network of cocited references derived from 5135 articles published on small-world networks between 1990 and 2010. The network of 205 references and 1164 cocitation links is divided into 12 clusters with a modularity of 0.6537 and the mean silhouette of 0.811. The red lines are made by the top-15 articles with the largest centrality variation rate.

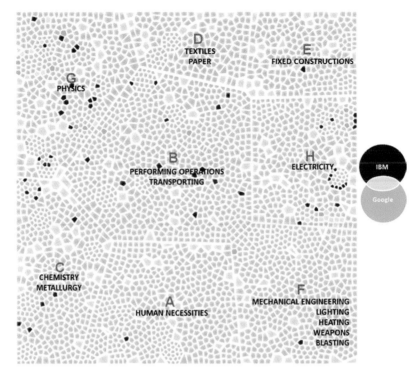

FIGURE 6.8 A Voronoi diagram of patents by IPC. Red, IBM; green, Google; yellow, both.

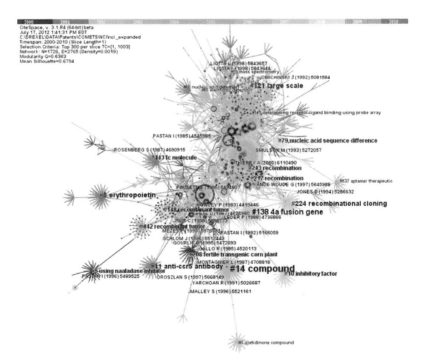

FIGURE 6.9 A minimum spanning tree of a network of 1726 cocited patents related to cancer research.

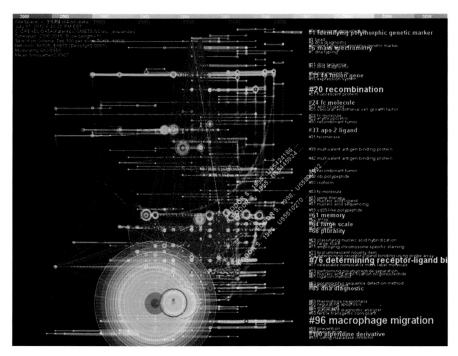

FIGURE 6.10 A timeline visualization of a broader context of the NCI patents. Dashed lines represent transformative links connecting different clusters of patents. Patents in each cluster are shown horizontally by their granted date from left to right. Labels next to clusters on the right characterize the primary topics of impact.

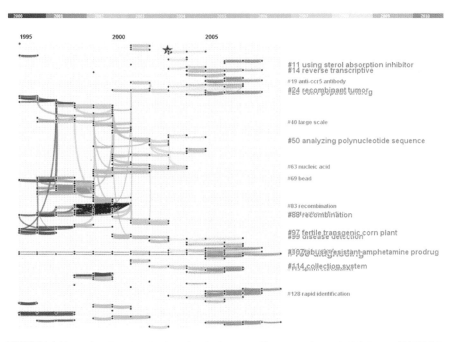

FIGURE 6.14 A timeline visualization of cocited patents. The star on the top is U.S. Patent 6537746. It contributed novel links connecting clusters from 83 through 88.

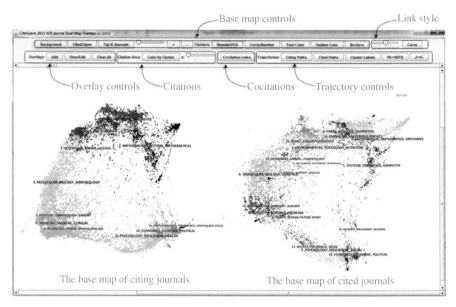

FIGURE 7.6 The initial appearance of the Dual-Map user interface, showing both citing and cited journal base maps simultaneously. The base map of 10,330 citing journals is on the left. The base map of 10,253 cited journals is on the right. The colors depict clusters identified by the Blondel clustering algorithm.

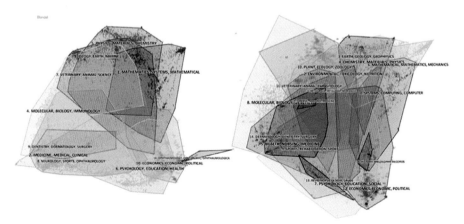

FIGURE 7.7 The boundary of each cluster is shown to depict how its members are distributed. Clusters in both base maps overlap substantially.

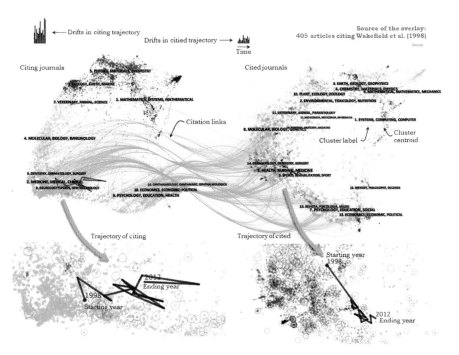

FIGURE 7.8 Citation patterns in an overlay of 405 articles that cited the Wakefield paper.

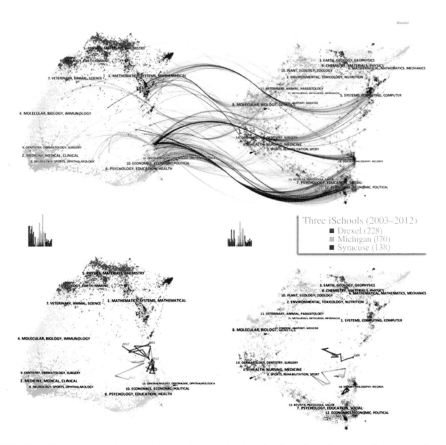

FIGURE 7.9 Overlays of three iSchools show major threads of citations that may characterize the publication portfolios of these institutions. The lower half of the figure shows the citing and cited trajectories in each of the base maps.

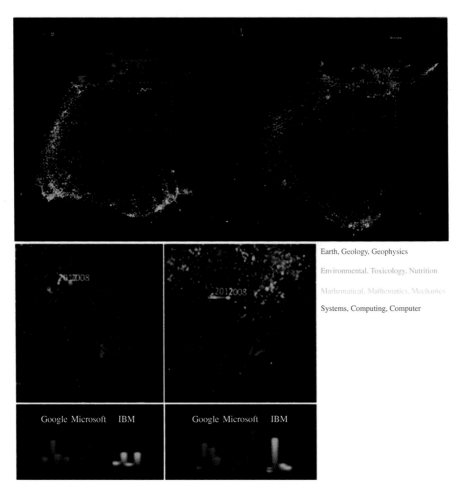

FIGURE 7.10   Trajectories of Google (blue), Microsoft (red), and IBM (green).

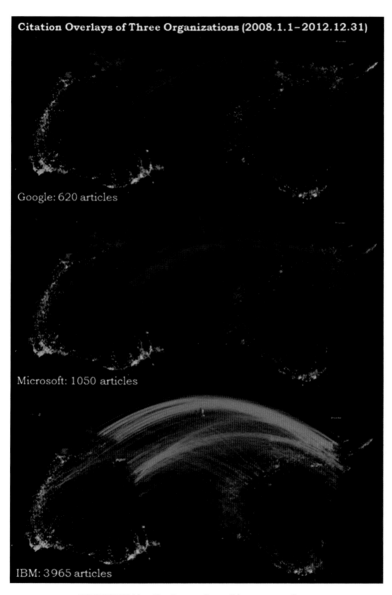

FIGURE 7.11 Citation overlays of three corporations.

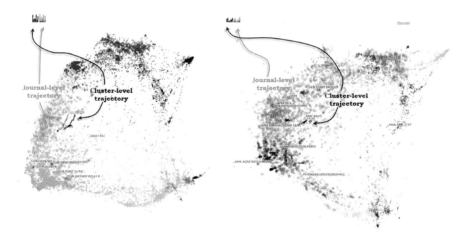

FIGURE 7.12 Trajectories of regenerative medicine research (2005–2012). The citing trajectory remains to be in the disciplinary area labeled as molecular, biology, and immunology throughout the entire course.

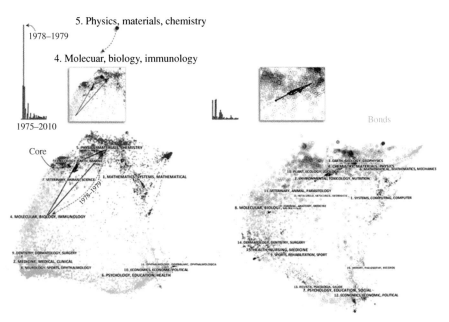

FIGURE 7.13 Trajectories of research in mass extinctions (1975–2010) at the discipline level. The core discipline of the research is identified as the Blondel cluster 3 on ecology, earth, and marine. The longest single-year shift occurred between 1978 and 1979 as the disciplinary center of the journals shifted from the Blondel cluster 5 on physics, materials, and chemistry to the Blondel cluster 4 on molecular, biology, and immunology.

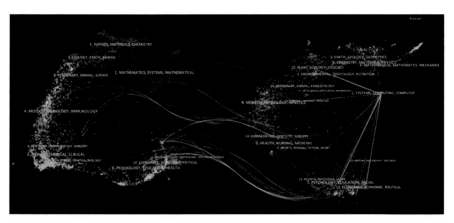

FIGURE 7.14 An overlay of publications in visual analytics (2006–2012). Wavelike curves depict citation links. They are colored by their source clusters. Dashed lines depict cocitation links across disciplinary boundaries.

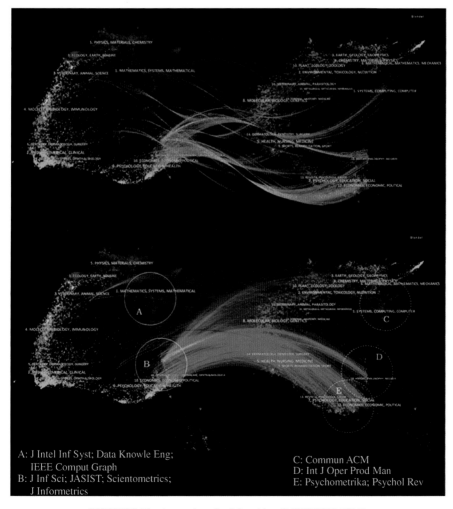

A: J Intel Inf Syst; Data Knowle Eng;
  IEEE Comput Graph
B: J Inf Sci; JASIST; Scientometrics;
  J Informetrics

C: Commun ACM
D: Int J Oper Prod Man
E: Psychometrika; Psychol Rev

FIGURE 7.15 An overlay of articles citing *JASIST* (2002–2011).

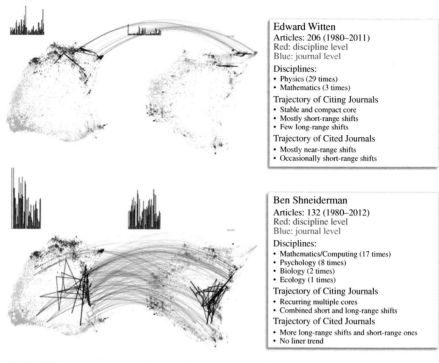

Edward Witten
Articles: 206 (1980–2011)
Red: discipline level
Blue: journal level

Disciplines:
• Physics (29 times)
• Mathematics (3 times)

Trajectory of Citing Journals
• Stable and compact core
• Mostly short-range shifts
• Few long-range shifts

Trajectory of Cited Journals
• Mostly near-range shifts
• Occasionally short-range shifts

Ben Shneiderman
Articles: 132 (1980–2012)
Red: discipline level
Blue: journal level

Disciplines:
• Mathematics/Computing (17 times)
• Psychology (8 times)
• Biology (2 times)
• Ecology (1 times)

Trajectory of Citing Journals
• Recurring multiple cores
• Combined short and long-range shifts

Trajectory of Cited Journals
• More long-range shifts and short-range ones
• No liner trend

FIGURE 7.16 Characteristics of trajectories of Witten's publications and Shneiderman's publications. Citing trajectories of overlays are shown on the left. Cited trajectories are shown on the right.

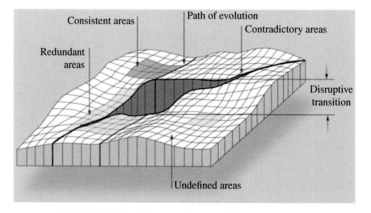

FIGURE 7.17 A fitness landscape of scientific inquires.

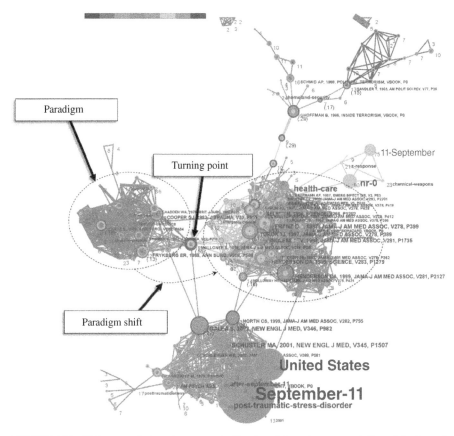

FIGURE 4.11 Major areas in terrorism research. (*See insert for color representation of the figure.*)
Source: Chen (2006).

on physical injuries resulted from terrorist attacks in the early 1990s. The cluster on
the right was formed by articles on the preparedness of health care concerning the
threats of biological and chemical weapons. The transition from the physical injury
cluster to the preparedness cluster was characterized by a single article labeled as the
turning point in the visualization. It is worth noting that the view of the transition
linkage between the two clusters might not exist at the level of individual researchers
if no one has ever made a citation chain that contains at least one article in each of
the clusters and the turning point article. It is quite possible that researchers working
on each cluster do not know any articles in the other cluster, except the common
turning point article. In other words, synthesized visualizations like this can reveal
something that no individual would be able to see otherwise.

The label "Turning point" in the figure points to the node with a purple ring,
which indicates that its betweenness centrality is high. It is evident that the node
bridges the green cluster and the yellow/orange-colored cluster. The cluster at the
bottom of the visualization contains several prominent terms such as *September-11*,
United States, and *post-traumatic stress disorder*. In fact, Galea et al.'s article we

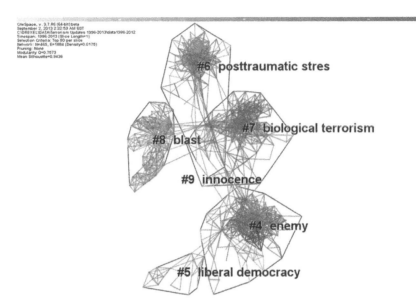

FIGURE 4.12   The simplicity of the research field at the cluster level.

mentioned earlier is located at the top of the cluster with a direct link with the turning point node. The visualized structure of the network has considerably reduced the complexity of the original dataset. It becomes easy to identify the major areas of terrorism research, how these areas are connected, and which articles have special positions. The significance of Galea et al.'s study is now conveyed by its role in linking the major areas. Furthermore, one does not need to have a substantial degree of domain expertise to notice the significance, even though we may not know specifically in what way it was special.

The next visualization shown in Figure 4.12, generated by a subsequently refined procedure, reduces the complexity even further. The entire network of 465 cocited references is divided into clusters. The label of each cluster indicates the context in which the cluster has been cited. The cluster no. 6, for example, labeled as "posttraumatic stres(s)," is where Galea et al.'s is located. Since the label was chosen from subsequently published articles that cited members of the cluster, it becomes possible for us to trace the citing articles and pinpoint exactly how they referred to Galea et al.'s study.

The next visualization in Figure 4.13 was generated by dropping the water level down even further so that we reveal more information about the metaphorical iceberg. The number of cocited references in the visualized network increased to 3638 by sampling top 500 relevant publications every year between 1996 and 2003. The structure of the network resembles the structure of the previous network generated by a smaller sample. The Galea et al.'s study published in the *New England Journal of Medicine* appears in the mid-right part of the image. It has a red core, which means there was a burst of citations in the first few years of its publication.

FIGURE 4.13   A visualization of a 3638-node network generated by sampling top 500 most cited articles each year between 1996 and 2003. (*See insert for color representation of the figure.*)

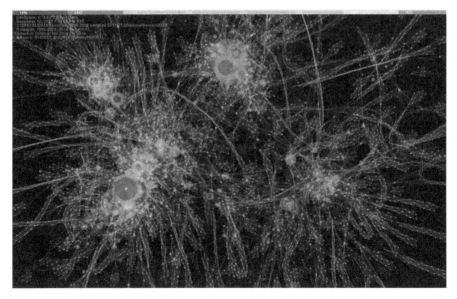

FIGURE 4.14   A network of 12,691 cocited references generated by sampling top 2000 most cited articles per year. (*See insert for color representation of the figure.*)

The final visualization of the terrorism case (Figure 4.14) shows a network of 12,591 cited references obtained by sampling 2000 most citing articles that are most cited every year between 1996 and 2003. The network contains 53,573 cocitation links. The three-cluster pattern is still evident at this level of granularity, which suggests that the simpler versions are reliable representations of the most salient structural properties.

4.3.2 Mass Extinctions

The study of mass extinctions is another interesting case. There have been five major mass extinctions on earth in the past. Although there have been numerous competing theories of what had caused these mass extinctions, the research problems to be addressed require a diverse range of expertise from disciplines such as geology, chemistry, paleontology, oceanography, and astronomy. Furthermore, since the most recent mass extinction took place over 65 million years ago, finding diagnostic evidence has been really challenging.

The initial visualization of the case revealed two major threads of research (see Figure 4.15). The one labeled as KTB (65 Ma), referring to the Cretaceous-Tertiary boundary, started from the lower left and came to an end around 1993. The majority cocitation links in this thread were made in the early 1990s. The other thread, shown near the top of the chart and labeled as PTB (250 Ma), referring to the Permian-Triassic Boundary, emerged in the early 1990s. It appears that much of the research activity had shifted from the KTB thread to the PTB thread through an article with a high betweenness centrality—a purple ring.

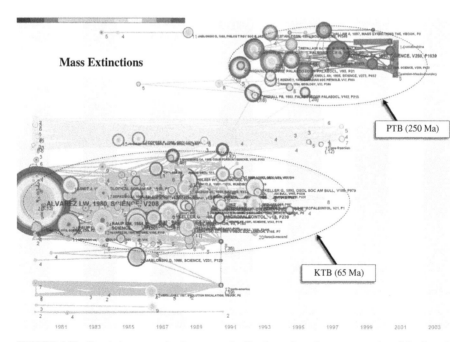

FIGURE 4.15 Trends in mass extinctions research. (*See insert for color representation of the figure.*) Source: Chen (2006).

Chen, C. (2006) pp. 369

comparable to that of the Chicxulub crater to the K-T impact theory. The discovery of the Chicxulub crater dramatically boosted the credibility of the K-T impact theory. Encouraged by the successful puzzle-solving experience, many scientists appear to have adapted the same approach to solve a different puzzle—by applying the impact theory to an earlier mass extinction. Finding the impact crater is the next logical step. Identifying a Permian-Triassic boundary impact crater has attracted the attention of many researchers. It was in this context that the current research front has emerged.

French B. M. and Keoberl C. (2010) pp. 152

The end of the Permian period, about 250 Ma ago, is marked by the largest known mass extinction in geological history. At this time, in two closely-separated events, more than the 90% of known marine species disappeared, accompanied by a major portion of terrestrial species as well (Erwin, 1993, 2006). Since the establishment of a firm connection between the later K–T extinction and a major impact event (Alvarez et al., 1980), numerous workers have searched for evidence of a similar connection between another large impact event and the Permian extinctions. Most efforts have concentrated on the younger and larger of the two extinction events, which marks the actual Permian–Triassic (P–Tr) boundary at 251 Ma.

FIGURE 4.16 The patterns identified in our 2006 article were found 4 years later by experts of mass extinctions in 2010.

The KTB thread is marked with a theory, the large purple circle. It was a 1980 theory that hypothesized that the K–T boundary mass extinctions were caused by asteroids smashing onto the earth. There were over 80 competing theories about the same time. No diagnostic evidence was found until the beginning of the 1990s. The success of the KTB theory clearly inspired the subsequent pursuit of the PTD thread, with a similar theory as its starting point.

We reported this system-level pattern in an article published in 2006. Intriguingly, as shown in Figure 4.16, an article published in 2010 by mass extinction experts reached an almost identical conclusion about the shift of research focus (French & Koeberl, 2010). This is particularly encouraging—we were able to identify the same pattern purely based on progressive knowledge domain analysis despite the fact that we had no domain expertise in the subject area. In another study published in May 2012, we applied the method to a computationally guided systematic review of the field of regenerative medicine. We were able to identify not only the rapidly advancing area of research on induced pluripotent stem cells (iPSCs) but also key publications that revolutionized the iPSC research. Five months later, in October 2012, the Nobel Committee announced that the Nobel Prize in Physiology or Medicine was awarded jointly to Sir John B. Gurdon and Shinya Yamanaka "for the discovery that mature cells can be reprogrammed to become pluripotent." It was Yamanaka and his team who authored the two revolutionary articles.

In February 2011, we reran the study after updating the relevant articles in the field (see Figure 4.17). As shown in the visualization of the domain, the largest cluster no. 36 permian–triassic boundary clearly represents the PTB thread we identified in our 2006 study. It confirms that our interpretation of the 2006 visualization turned out to be accurate. The predominant thread of research is the PTB thread.

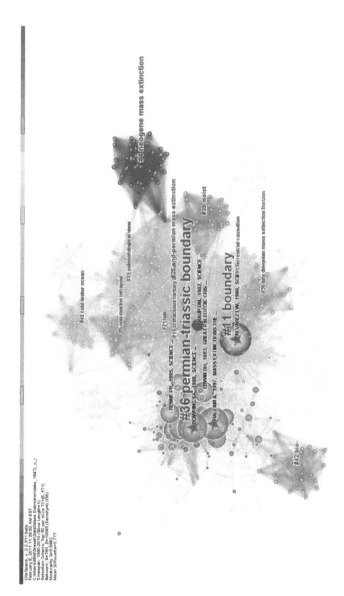

FIGURE 4.17   A visualization of mass extinction in a follow-up study in 2011. (*See insert for color representation of the figure.*)

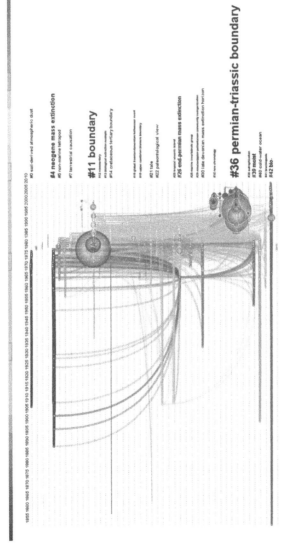

FIGURE 4.18 A timeline visualization of mass extinctions research.

Figure 4.18 shows a timeline visualization of mass extinctions research with the data updated in 2011. Cluster no. 36 corresponds to the PTB thread in the 2006 visualization. Cluster no. 11 corresponds to the KTB thread, including the highly cited impact theory article, shown as the largest circle in this thread. Although the new version shows more details and more threads, the two most prominent threads resemble the visualization in the 2006 publication.

4.3.3   Developing Expertise in Analytics and Topic Areas

CiteSpace is widely used for visualizing and analyzing patterns and trends in scientific literature. Every day there are hundreds of individual users or more from different countries. Some of these users are beginners, whereas others are more experienced. Presently, concurrent users are invisible to each other, thus they have no way to communicate and discuss with other users who may have experienced the same issues. Since users are engaged in tasks such as analyzing and interpreting visualized networks, they often find themselves in need to consult and discuss their situations with someone with more experience. At a higher level, the currently invisible community represents a potentially valuable insight of what people are interested in terms of which part of scientific literature they choose to analyze.

Despite the growing number of tools available to users, the complexity involved in using such tools and, more importantly, in analyzing the results produced by these tools, has been an increasing challenge to users as well as designers of such tools. Having realized the challenge, designers and developers of the tools have been running tutorials and workshops in an attempt to reduce the burden on users. Nevertheless, new users have been attracted to the new ways of studying scientific literature at a much faster rate than what tutorials and workshops can accommodate. Furthermore, most of the tools have been actively maintained and frequently upgraded with new features and more powerful capabilities. As an analyst, users often need to make numerous decisions as they try to find their way through a potentially large decision space because multiple routes may be available. Although multiple choices provide a level of flexibility for experienced users, a novice user may find it challenging to deal with the complexity without extra help. Users need to keep up with the advance of tools. They need to develop their analytic skills as they study a new topic. They are in need of instant feedback on how they should interpret what is being revealed to them by an enabling tool and what they can make of it.

Few visual analytic systems in the current generation, if any, address users' needs for just-in-time learning in terms of operating a visual analytic tool and facilitating users to derive insights into the body of scientific literature that they are investigating. Reducing the complexity involved in learning and using this type of visual analytic tools have profound implications. CiteSpace has attracted an active user population of a considerable size with users from many countries. For example, over 800 distinct users used CiteSpace in the month of August 2013. Figure 4.19 shows a visualization of the distribution of CiteSpace's users. Users' geographic locations are aggregated at the city level. The height of a bar is proportional to the average size of the networks that have been analyzed in CiteSpace.

FIGURE 4.19 The distribution of CiteSpace's users. The height of a bar represents the average size of networks analyzed by users at a city. (*See insert for color representation of the figure.*)

freq ▲	ip	cluster_label
23	222.66.	new economic geography
17	218.18.	critical question
15	117.90.	scoring function
14	111.174.	comparative literature
13	60.2.	bioterrorism
12	117.90.	capability
12	58.51.	贯彻中央思想政治工作会议精神
11	58.51.	"思想政治工作";"宣传工作";"精神文明建设"
9	180.118.	synthesis
9	117.90.	scoring function
9	183.157.	innovation
8	58.51.	"精神文明";"先进文化";"三个代表";"中国特色社会主义"
8	221.232.	bioterrorism
8	218.57.	diesel engine
8	183.62.	从"广誉远"看中医药非物质文化遗产保护的价值
7	187.114.	performance measurement
7	143.107.	solar cell
7	186.207.	poly
7	202.196.	"留守儿童";"研究范式";"问题分析"
7	121.232.	drug design

FIGURE 4.20 The topics of the largest clusters in networks analyzed by CiteSpace users grouped by IP locations.

Our preliminary study has shown that users often start with relatively small-sized networks and explore various analytic and visualization features before they move on to analyze larger networks. Thus, the average size of networks may provide a coarse indicator of expertise of using CiteSpace. A potential application that could make use of this type of information is an expert locator.

Figure 4.20 illustrates the labels of the largest clusters in networks analyzed by users grouped by their IP locations. For example, a user at 222.66.■■.■ analyzed the literature of new economic geography. A user at 60.2.■■.■ analyzed a dataset with the largest cluster on bioterrorism, which indicates that the user was probably exploring with the built-in dataset. It is possible to combine the distribution of expertise in the analytics process with the distribution of topics within a common time-frame, such as the last hour or last 7 days, so that one can find users with expertise either in the analytic procedure or in the topic areas, or both.

4.3.4   U.S. Supreme Court Landmark Cases

The same analytic procedure can be applied to U.S. Supreme Court Opinions and U.S. patents because both of them include citations that are similar, at least to computer algorithms, to citations in scientific literature. Figure 4.21 illustrates a visualization of landmark cases of the U.S. Supreme Court Opinions. Dark inner circles such as

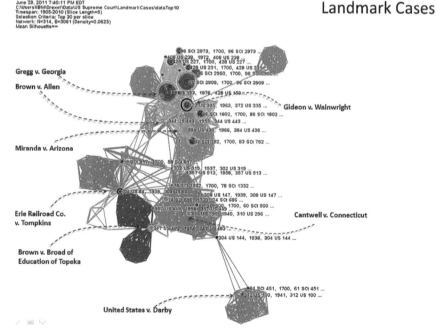

FIGURE 4.21   A visualization of U.S. Supreme Court landmark cases.

the one for Gregg v. Georgia near the top represent cases that have citation bursts—abrupt increases in citations made by subsequent Supreme Court decisions. The same interactions and analytic tasks can be applied and interpreted in the new context. Cases with solid black rings such as the one with Gideon v. Wainwright indicate possible Gestalt switch cases in the legal landscape. They may set a new precedent for subsequent cases. Analyzing specific ways that a case has been cited may identify valuable insights into the role of the case and whether its role has been drifting or has abruptly changed.

4.4 SUMMARY

In this chapter, we introduce a generic approach to the study of a knowledge domain. We have demonstrated the approach based on structural patterns that can be derived from citation patterns and how such patterns evolve over time. Several examples are used to illustrate the basic procedure and the types of questions one can address using this approach. The value of this approach is that it can often reduce the complexity of an underlying phenomenon to a level of clarity that can bring us insights into the critical structural and dynamic properties and patterns.

BIBLIOGRAPHY

Antonakis, J., & Lalive, R. (2008). Quantifying scholarly impact: IQp versus the Hirsch *Journal of the American Society for Information Science and Technology, 59*(6), 956–969.

Bonacich, P. (1987). Power and centrality: A family of measures. *American Journal of Sociology, 92*, 1170–1182.

Brandes, U. (2001). A faster algorithm for betweenness centrality. *Journal of Mathematical Sociology, 25*(2), 163–177.

Brody, T., Harnad, S., & Carr, L. (2006). Earlier web usage statistics as predictors of later citation impact. *Journal of the American Association for Information Science and Technology, 57*(8), 1060–1072.

Burt, R. S. (1992). *Structural holes: The social structure of competition*. Cambridge, MA: Harvard University Press.

Burt, R. S. (2004). Structural holes and good ideas. *American Journal of Sociology, 110*(2), 349–399.

Burt, R. S. (2005). *Brokerage and closure: An introduction to social capital*. Oxford, UK: Oxford University Press.

Chen, C. (2004). Searching for intellectual turning points: Progressive knowledge domain visualization. *Proceedings of the National Academy of Sciences of the United States of America, 101*(Suppl.), 5303–5310.

Chen, C. (2005). *The centrality of pivotal points in the evolution of scientific networks*. Paper presented at the International Conference on Intelligent User Interfaces (IUI 2005), January 10–13, 2005, San Diego, CA.

Chen, C. (2006). CiteSpace II: Detecting and visualizing emerging trends and transient patterns in scientific literature. *Journal of the American Society for Information Science and Technology, 57*(3), 359–377.

Chen, C., Chen, Y., Horowitz, M., Hou, H., Liu, Z., & Pellegrino, D. (2009). Towards an explanatory and computational theory of scientific discovery. *Journal of Informetrics, 3*(3), 191–209.

Chen, C., Hu, Z., Liu, S., & Tseng, H. (2012). Emerging trends in regenerative medicine: A scientometric analysis in CiteSpace. *Expert Opinions on Biological Therapy, 12*(5), 593–608.

Chen, C., Ibekwe-SanJuan, F., & Hou, J. (2010). The structure and dynamics of co-citation clusters: A multiple-perspective co-citation analysis. *Journal of the American Society for Information Science and Technology, 61*(7), 1386–1409.

Chen, C., & Kuljis, J. (2003). The rising landscape: A visual exploration of superstring revolutions in physics. *Journal of the American Society for Information Science and Technology, 54*(5), 435–446.

Chen, C., Zhang, J., & Vogeley, M. S. (2009). Mapping the global impact of Sloan Digital Sky Survey. *IEEE Intelligent Systems, 24*(4), 74–77.

Crane, D. (1972). *Invisible Colleges: Diffusion of Knowledge in Scientific Communities.* Chicago: University of Chicago Press.

Cronin, B. (1981). Agreement and divergence on referencing practice. *Journal of Information Science, 3*(1), 27–33.

Dalen, H. P. v., & Henkens, K. (2005). Signals in science – On the importance of signaling in gaining attention in science. *Scientometrics, 64*(2), 209–233.

Deerwester, S., Dumais, S. T., Landauer, T. K., Furnas, G. W., & Harshman, R. A. (1990). Indexing by latent semantic analysis. *Journal of the American Society for Information Science, 41*(6), 391–407.

Dunbar, K. (1993). Concept discovery in a scientific domain. *Cognitive Science, 17,* 397–434.

Dunning, T. (1993). Accurate methods for the statistics of surprise and coincidence. *Computational Linguistics, 19*(1), 61–74.

Fiszman, M., Demner-Fushman, D., Kilicoglu, H., & Rindflesch, T. C. (2009). Automatic summarization of MEDLINE citations for evidence-based medical treatment: A topic-oriented evaluation. *Journal of Biomedical Informatics, 42,* 801–813.

Freeman, L. C. (1977). A set of measuring centrality based on betweenness. *Sociometry, 40,* 35–41.

French, B. M., & Koeberl, C. (2010). The convincing identification of terrestrial meteorite impact structures: What works, what doesn't, and why. *Earth-Science Reviews, 98,* 123–170.

Galea, S., Ahern, J., Resnick, H., Kilpatrick, D., Bucuvalas, M., Gold, J., & Vlahov, D. (2002). Psychological sequelae of the September 11 terrorist attacks in New York City. *New England Journal of Medicine, 346*(13), 982–987.

Garfield, E. (1979). *Citation Indexing: Its Theory and Applications in Science, Technology, and Humanities.* New York: John Wiley.

Garfield, E., & Welljamsdorof, A. (1992). Of nobel class: A citation perspective on high-impact research authors. *Theoretical Medicine, 13*(2), 117–135.

Guiffre, K. (1999). Sandpiles of opportunity: Success in the art world. *Social Forces, 77*(3), 815–832.

Heinze, T., & Bauer, G. (2007). Characterizing creative scientists in nano-S&T: Productivity, multidisciplinarity, and network brokerage in a longitudinal perspective. *Scientometrics, 70*(3), 811–830.

Hirsch, J. E. (2005). An index to quantify an individual's scientific output. *Proceedings of the National Academy of Sciences of the United States of America, 102,* 16569.

Holton, G., Chang, H., & Jurkowitz, E. (1996). How a scientific discovery is made: A case history. *American Scientist, 84*(4), 364–365.

Jaccard, P. (1901). Étude comparative de la distribution florale dans une portion des Alpes et des Jura. *Bulletin del la Société Vaudoise des Sciences Naturelles, 37,* 547–579.

Kamada, T., & Kawai, S. (1989). An algorithm for drawing general undirected graphs. *Information Processing Letters, 31*(1), 7–15.

Kessler, M. M. (1963). Bibliographic coupling between scientific papers. *Journal of the American Society for Information, 14*(1), 10–25.

Kiss, C., & Bichler, M. (2008). Identification of influencers: Measuring influence in customer networks. *Decision Support Systems, 46*(1), 233–253.

Kleinberg, J. (2002). *Bursty and hierarchical structure in streams.* Paper presented at the proceedings of the 8th ACM SIGKDD international conference on Knowledge Discovery and Data Mining, July 23–26, 2002, Edmonton, Alberta. New York: ACM Press. Retrieved June 15, 2010, from http://www.cs.cornell.edu/home/kleinber/bhs.pdf. Accessed on February 21, 2014.

Kuhn, T. S. (1962). *The Structure of Scientific Revolutions.* Chicago: University of Chicago Press.

Kuhn, T. S. (1970). *The Structure of Scientific Revolutions* (2nd ed.). Chicago: University of Chicago Press.

Kumar, R., Novak, J., Raghavan, P., & Tomkins, A. (2003). *On the bursty evolution of blog-space.* Paper presented at the WWW2003. Retrieved June 1, 2010, from http://www2003.org/cdrom/papers/refereed/p477/p477-kumar/p477-kumar.htm. Accessed on February 21, 2014.

Laudan, L., Donovan, A., Laudan, R., Barker, P., Brown, H., Leplin, J., et al. (1986). Scientific change: Philosophical models and historical research. *Synthese, 69*(2), 141–223.

Lokker, C., McKibbon, K. A., McKinlay, R. J., Wilczynski, N. L., & Haynes, R. B. (2008). Prediction of citation counts for clinical articles at two years using data available within three weeks of publication: Retrospective cohort study. *British Medical Journal, 336*(7645), 655–657.

Luxburg, U. v. (2006). *A tutorial on spectral clustering.* Retrieved June 15, 2010, from http://www.kyb.mpg.de/fileadmin/user_upload/files/publications/attachments/Luxburg07_tutorial_4488%5b0%5d.pdf. Accessed on February 21, 2014.

Luxburg, U. v., Bousquet, O., & Belkin, M. (2009). *Limits of spectral clustering.* Retrieved June 15, 2010, from http://kyb.mpg.de/publications/pdfs/pdf2775.pdf. Accessed on February 21, 2014.

Newman, M. E. J. (2006). Modularity and community structure in networks. *Proceedings of the National Academy of Sciences of the United States of America, 103*(23), 8577–8582.

Nikkila, J., Toronen, P., Kaski, S., Venna, J., Castren, E., & Wong, G. (2002). Analysis and visualization of gene expression data using self-organizing maps. *Neural Networks, 15*(8–9), 953–966.

Perneger, T. V. (2004). Relation between online "hit counts" and subsequent citations: Prospective study of research papers in the BMJ. *British Medical Journal, 329*, 546–547.

Radder, H. (1997). Philosophy and history of science: Beyond the Kuhnian paradigm. *Studies in History and Philosophy of Science, 28*(4), 633–655.

Rousseeuw, P. J. (1987). Silhouettes: A graphical aid to the interpretation and validation of cluster analysis. *Journal of Computational and Applied Mathematics, 20*, 53–65.

Salton, G., Yang, C. S., & Wong, A. (1975). A vector space model for information retrieval. *Communications of the ACM, 18*(11), 613–620.

Schvaneveldt, R. W. (Ed.). (1990). *Pathfinder Associative Networks: Studies in Knowledge Organization*. Norwood, NJ: Ablex Publishing Corporations.

Shi, J., & Malik, J. (2000). Normalized cuts and image segmentation. *IEEE Transactions on Pattern Analysis and Machine Intelligence, 22*(8), 888–905.

Shibata, N., Kajikawa, Y., Taked, Y., & Matsushima, K. (2008). Detecting emerging research fronts based on topological measures in citation networks of scientific publications. *Technovation, 28*(11), 758–775.

Small, H. (1973). Co-citation in the scientific literature: A new measure of the relationship between two documents. *Journal of the American Society for Information Science, 24*, 265–269.

Small, H. (2000). Charting pathways through science: Exploring Garfield's vision of a unified index to science. In B. Cronin & H. B. Atkins (Eds.), *The web of knowledge: A festschrift in honor of eugene garfield* (pp. 449–473). Medford, NJ: Information Today Inc.

Small, H. G. (1977). A co-citation model of a scientific specialty: A longitudinal study of collagen research. *Social Studies of Science, 7*, 139–166.

Sparck-Jones, K. (1999). Automatic summarizing: Factors and directions. In I. Mani & M. T. Maybury (Eds.), *Advances in automatic text summarization* (pp. 2–12). Cambridge, MA: MIT Press.

Swanson, D. R. (1986). Fish oil, Raynaud's syndrome, and undiscovered public knowledge. *Perspectives in Biology and Medicine, 30*, 7–18.

Teufel, S., & Moens, M. (2002). Summarizing scientific articles: Experiments with relevance and rhetorical status. *Computational Linguistics, 28*(4), 409–445.

White, H. D. (2007a). Combining bibliometrics, information retrieval, and relevance theory, Part 1: First examples of a synthesis. *Journal of the American Society for Information Science and Technology, 58*(4), 536–559.

White, H. D. (2007b). Combining bibliometrics, information retrieval, and relevance theory, Part 2: Some implications for information science. *Journal of the American Society for Information Science and Technology, 58*(4), 583–605.

Wiener, N. (1948). *Cybernetics or control and communication in the animal and the machine*. Cambridge, MA: MIT Press.

CHAPTER 5

Fitness Landscapes

5.1 COGNITIVE MAPS

The origin of cognitive maps can be traced back to Edward Tolman's famous study of how rats were able to find the food in a maze[1] (Tolman, 1948). Tolman discovered that rats had obviously memorized the layout of the maze. Prior to Tolman's study, rats were not believed to be capable of learning the layout of a maze. Tolman called the internalized layout a cognitive map. He further proposed that rats and other organisms are able to develop cognitive maps of their environments. For example, our mental image of the layout of a city is developed in a similar way.

People's cognitive maps may be different, even if they live in the same environment. A commonly used method to study people's internal cognitive maps is to ask them to sketch their mental maps on paper. Geographers often ask people to sketch a map of an area with directions to a landmark or a specific location. They may also ask people to give the names of many places as quickly as they can so as to reveal how strongly two places are associated in one's cognitive map.

5.1.1 The Legibility of Cognitive Maps

Kevin Lynch (1960) attempted to restore the social and symbolic function of public spaces and make modern cities "legible." In *The Image of the City*, Lynch argued that we need to know where we are in a city and have a mental image of each part of the city. In particular, the legibility of a city is defined by people's *mental maps*.

The legibility of a place is how easily one can understand its layout. Lynch suggested five elements of legibility: paths, edges, districts, nodes, and landmarks.

---

[1]http://psychclassics.yorku.ca/Tolman/Maps/maps.htm

The Fitness of Information: Quantitative Assessments of Critical Evidence, First Edition. Chaomei Chen.
© 2014 John Wiley & Sons, Inc. Published 2014 by John Wiley & Sons, Inc.

Paths are familiar routes that people use to move from one location to another. A city has a network of major routes and a network of minor routes.

Districts are areas with perceived internal homogeneity. They are medium-to-large sections of the city. They share common characteristics.

Edges are paths that separate districts.

Nodes are attraction centers in a city, such as the Trafalgar Square in London and Times Square in New York.

Landmarks are visually prominent points of reference in a city, such as the Statue of Liberty in New York, the Big Ben in London, and the Eiffel Tower in Paris. A landmark is a visually prominent object, whereas a node is a hub of activity.

In *The Image of the City*, Lynch described his studies of three American cities. Manhattan, for example, has a grid layout. Travelers who are aware of this layout can guide themselves with this information. Rob Ingram and Steve Benford incorporated this type of legibility features into the design of an information visualization system of a collection of document so as to improve its legibility (Ingram & Benford, 1995). They simplified the layout of a graph representation of interrelated documents using Zahn's minimum spanning tree algorithm (Zahn, 1917). In addition, they removed some of the weakest links so that the resultant minimum spanning tree was divided further into a number of smaller trees. These trees correspond to the idea of a district in a city. Ingram and Benford also included landmarks in their displays. If three districts join together in the information space, then the centers of these districts form a triangle. A landmark can be added at the center of the triangle. Edges are drawn to highlight the boundaries of large districts.

5.1.2 Spatial Knowledge

The legibility of a city helps people find their ways in the city. The more spatial knowledge we have of a city, the easier we can navigate it. Thorndyke and Hayes-Roth distinguished three levels of such spatial knowledge (Thorndyke & Hayes-Roth, 1982) in terms of landmark knowledge, procedural knowledge, and survey knowledge.

Landmark knowledge is the most basic level of awareness of the layout of a place or an abstract environment. If all we know about London is the Big Ben, then our ability to navigate through London would be rather limited.

Procedural knowledge, also known as route knowledge, enables a traveler follow a particular route from one location to another. Procedural knowledge is more powerful than landmark knowledge because it expands landmark knowledge to a larger and more flexible network for a traveler. At this level, we should be able to follow at least one route that can take us from the Big Ben to the Trafalgar Square.

At the highest level, we have obtained *survey knowledge* of a city or a conceptual environment. We have internalized the layout of a city to such an extent that we can find alternative routes between two locations whenever we want. Survey knowledge is essential for us to explore and make discoveries.

Everyone applying for a taxi license in London must pass the Knowledge of London exam. Each applicant needs to demonstrate that he or she has a thorough knowledge of London. The applicant would need to know the locations of various places in London. In short, the applicant needs to know the best way to get the passengers to their destinations.

The best way to acquire survey knowledge is from a hands-on experience in an environment. Alternatively, we can develop our survey knowledge by doing much homework and reading many maps. Survey knowledge acquired in this way, however, is orientation specific, which means that the navigator may need to rotate the mental representation of the space to match the environment. This concern led Marchon Levine to explore how this phenomenon should be taken into account by map designers (Levine, Jankovic, & Palij, 1982; Levine, Marchon, & Hanley, 1984). Levine argued that maps should be congruent and synchronized with the environment so that we can quickly match our current position with the map. Levine proposed three principles for map design:

1. *The two-point theorem*—a map reader must be able to relate two points on the map to their corresponding two points in the environment.
2. *The alignment principle*—the map should be aligned with the terrain. A line between any two points in space should be parallel to the line between those two points on the map.
3. *The forward-up principle*—the upward direction on a map must always show what is in front of the viewer.

Researchers have adapted real-world way-finding strategies for performing similar tasks in a virtual environment. Rudolph Darken and others provide an informative summary on way-finding behavior in virtual environments (Darken, Allard, & Achille, 1998). The legibility of the environment is essential for us in our search for what we need in such environments. It becomes even more significant if the environment is abstract, complex, and adaptive.

5.2   FITNESS LANDSCAPES

Evolution is a complex process. An intriguing metaphor that can help us grasp the complexity of a population's evolution is a fitness landscape. Fitness landscapes, adaptive landscapes, and evolutionary landscapes are terms frequently used to describe a group of similar ideas.

The original metaphor of a fitness landscape was proposed in the early 1930s by Sewell Wright. A fitness landscape represents the fitness or the likelihood of survival of an organism in terms of its genes and their combinations. Wright demonstrated that this metaphor provides an intuitive vehicle to communicate a complex idea. Researchers from a diverse range of disciplines have been inspired by the notion of a fitness landscape and developed a variety of applications. On the other hand, the concept of a fitness landscape has received criticisms from philosophers, biologists, and others. It is our view that the idea is definitely inspirational and that it is the responsibility of its adopters to ensure a valid use and interpretation.

5.2.1 Wright's Adaptive Landscapes

Wright published the initial idea in 1931 on evolution in Mendelian populations. He presented the topic at the Sixth International Congress of Genetics in 1932. Wright was originally motivated to describe the ideal conditions for evolution to take place. With the ideal conditions, evolution would reach its highest adaptive peak at the fastest rate. He encountered a roadblock for explaining how populations may temporarily shift to a lower level of fitness because natural selection will always push the population to a higher level of fitness. He proposed a three-phase *shifting balance theory* to explain how evolution proceeds:

1. Random genetic drift causes subpopulations semi-isolated within the global population to lose fitness.
2. Selection on complex genetic interaction systems raises the fitness of those subpopulations.
3. Interdemic selection then raises the fitness of the large or global population.

Wright's adaptive landscapes are multidimensional models represented on two-dimensional depictions. The surface of the landscape represents the adaptive value assigned to the underlying gene combinations. Each peak would represent a point where the combinations of genes and the environment reach a state of harmony. In their recent chapter on "A Shifting Terrain," Michael R. Dietrich and Robert A. Skipper, Jr., presented a brief history of the adaptive landscape. Of particular interest is how Wright's original metaphor of an adaptive landscape at the genotype level has been expanded to phenotypic and molecular landscapes.

In his 1932 paper, Wright illustrated the dimensionality challenge. A species with 10^{1000} combinations requires 9000 dimensions. Although high-dimensional fitness landscapes are logically possible, the idea is often explained in a two- or three-dimensional form. Generally speaking, an n-dimensional fitness landscape can be defined by a mapping f from a domain of $n-1$ dimensions to a one-dimensional fitness dimension: $f: \mathbf{R}^{n-1} \rightarrow \mathbf{R}^1$. A point in an $n-1$ dimension space corresponds to a combination of $n-1$ attributes. The fitness or the survival probability of an organism is a function of such combinations.

Epistasis is a concept in genetics. Epistasis refers to the interaction between different genes when the expression of one gene depends on the presence of so-called modifier genes. The term epistatic was first used in 1909 to describe the effect that a variant or allele at one locus prevents the variant at another locus from manifesting its effect. The genetic roles of epistasis in complex diseases such as diabetes, asthma, hypertension, and multiple sclerosis have attracted much attention, although it appears to be very challenging due to many complicating factors such as a large number of contributing loci and susceptibility alleles. Heather J. Cordell, University of Cambridge, produced a survey of the topic in 2002.[2]

---

[2]http://hmg.oxfordjournals.org/content/11/20/2463.full

Suppose B and G are two loci that influence a trait such as hair color in mice. Locus B has two possible alleles B and b, and locus G has two possible alleles G and g. The hair color (white, black, or gray) is the phenotypic outcome of the underlying genotype. It is found that at locus G, allele G is dominant to allele g regardless of the genotype at locus B. In other words, the effect of allele g is masked by G.

Allele G at locus G is epistatic to allele B at locus B, which means that the effect at locus B is masked by that of locus G. Mathematically, the genetic concept of epistasis can be represented by a linear model as follows:

$$z \sim \sum \alpha_i x_i + \sum \beta_j y_j + \sum \gamma_{ij} x_i y_j$$

where z is a quantitative phenotype and $\{x_i\}$ and $\{y_j\}$ are variables related to the underlying genotype at loci i and j. The coefficients α_i and β_j represent genetic parameters that may be estimated corresponding to the mean effect and additive and dominance effects at the two loci. In contrast, γ_{ij} corresponds to epistatic interaction effects. If there is no epistasis, that is, no interaction, then the model can be simplified:

$$z \sim \sum \alpha_i x_i + \sum \beta_j y_j$$

The genotype of an organism is its inherited genetic instructions. A phenotype is an organism's observable features, such as brown eyes in humans. Phenotypes result from the gene expression of an organism and the influence of its environment. Thus, phenotypes are due to interactions between genes and environmental factors:

$$\text{Genotype}\,(G) + \text{environment}\,(E) +$$
$$\text{genotype \& environment interactions}\,(GE) \to \text{phenotype}\,(P)$$

A fundamental question regarding the basic properties of a fitness landscape is: what is it like? Does it have a peak? Or does it have multiple peaks of a wide variety? Does it have a smooth and continuous surface? What does it mean if a population moves across a valley between two peaks? What does it mean if we see two populations move in opposite directions? Figure 5.1 shows Wright's diagrams of adaptive landscapes. Contour lines depict the gradients of the landscapes.

Wright considered the paths different populations might take across the fitness landscape through random mutations and selection. For effective evolution to take place, a population must be able to move from a lower peak to a higher one. Intuitively, it seems that the population has to move across a saddle between the two peaks. It must accept a lower level of fitness as it moves across the saddle. Subsequently, this interpretation has been continuously criticized and defended.

In each scenario, a peak is marked by a plus sign, and a valley is marked by a minus sign. The dashed circle represents the initial position of the population on a peak (not necessarily the highest) in the fitness landscape. The shaded circle is the next position of the population after natural selection. The top row shows that since the population is so large, it will be largely unaffected by individual mutations. In the upper left

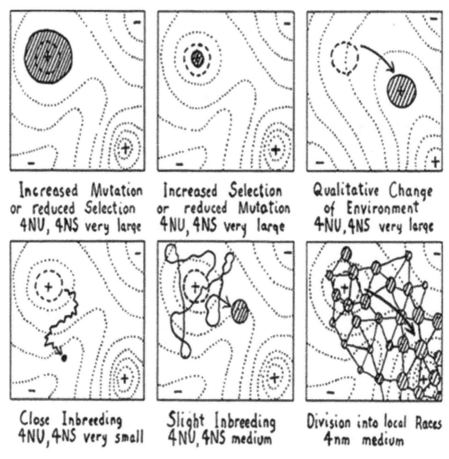

Increased Mutation | Increased Selection | Qualitative Change
or reduced Selection | or reduced Mutation | of Environment
4NU, 4NS very large | 4NU, 4NS very large | 4NU, 4NS very large

Close Inbreeding | Slight Inbreeding | Division into local Races
4NU, 4NS very small | 4NU, 4NS medium | 4nm medium

FIGURE 5.1 Sewall Wright's diagrams of adaptive landscapes. Each frame represents different evolutionary scenarios and their impact on the population. Frame C represents a dynamic landscape in a changing environment. Frame F represents the dynamics of Wright's shifting balance theory. Source: From Wright (1932).

diagram, the mutation rate is high. In other words, the selection pressure is weak, so the population remains around the fitness hilltop. In the top center diagram, the population concentrates at the fitness peak after living under unchanging conditions for a long time with either a low mutation rate or high selection pressure. In the top right diagram, the changing environment shifted the previous peak to a new location. As a result, the population follows the new fitness peak.

Diagrams in the second row show the dynamics of smaller populations. In the lower left diagram, the population is so small that it is affected by virtually every mutation. Consequently, it moves across the landscape randomly. In the bottom center diagram, the small population is affected nearly equally by mutations and selection pressure and easily escapes small peaks while moving very slowly up higher ones. The lower right diagram shows the optimal scenario from Wright's point of view: many small, isolated populations that occasionally interact. These groups can change quickly, but are spared the uncontrolled random wandering of a single small group by virtue of their interaction with one another.

If the surface of a fitness landscape is smooth, it means that a small change in the underlying combinations would not change much of the fitness value. On a rugged landscape, however, a small step away from the current position may lead to a dramatic change of the fitness value.

The earliest fitness landscapes were proposed to represent changes in fitness values driven by variations of gene combinations at the genotype level. Then, fitness landscapes were defined at the phenotype level and the molecular level. Theodosius Dobzhansky and G.G. Simpson were among the most influential researchers who played an instrumental role in the dissemination of the idea of adaptive landscapes. Dobzhansky included Wright's genetic landscapes in each edition of his *Genetics and the Origin of Species*. Dobzhansky's main contribution was considered to be the introduction of modern genetics to evolutionary synthesis in 1937.

In 1944, G.G. Simpson integrated paleontology with the adaptive landscape and added a new type of adaptive landscapes—the phenotypic landscape. Simpson brought paleontological evidence of evolution and how patterns of phenotypic change across species resulted from adaptation.

The third type of adaptive landscapes is at the molecular level as opposed to the genotype and phenotype levels. John Maynard Smith envisaged a network of proteins, and each protein is one mutational step away from the other. This metaphor leads to the question, "how it is possible for evolution to move from one functional protein to another by natural selection?" If natural selection is the only mechanism for evolution, then there must be a continuous path on the network, and the entire path should be made of functional proteins because a nonfunctional protein would be selected against. In the late 1960s, it was conceived that perhaps not all molecular structures are subject to natural selection. In other words, there may be selection-neutral proteins. The neutral theory drew much controversy over the role of natural selection at the molecular level. The neutral theory offered an alternative way to formulate the question about the existence of a continuous path of functional proteins. Now, the question is how often evolution has passed through a nonfunctional sequence of proteins. On an adaptive landscape, this means that a path of protein evolution may not move uphill all the time.

5.2.2 Fisher's Geometric Model of Adaptation

Ronald A. Fisher (1890–1962), a widely known statistician and evolutionary biologist, proposed a geometric model of evolution to characterize statistical properties of adaptation. The relationship between adaptation and fitness is similar to the relationship between a telescope with multiple lenses and the clarity of the image.

In essence, the design of a telescope is the configuration of a number of lenses placed along a straight line. There can be numerous possible configurations, depending on how many lenses are to be used, the optical properties of each lens, the physical properties of the lenses (such as concavity and size), and the distance between adjacent lenses. The quality of an individual configuration is the quality of the image that the telescope can produce. The quality of an image could be a multidimensional concept in its own right, including the resolution, contrast, and saturation of colors. For the sake of argument, suppose the quality of an image is measured by the sharpness of the image. One can change a configuration by making adjustments of one or more of the parameters mentioned earlier, for example, by swapping the positions of two adjacent lenses. As a result, the quality of the image may change for better or for worse, or it may not change at all. In the next chapter, we will discuss the design of a telescope from a different perspective, generating patentable ideas. Until then, we will use this example to illustrate the complex process of adaptation in evolutionary biology.

Each candidate for the configuration of a potentially useful telescope can be conceptualized as a point in a high-dimensional space. The quality of a configuration is a function of the position in the high-dimensional space. The goal of a design process is to find the position of a configuration that can produce images of the best possible quality.

Given that each step aside from the current position in the high-dimensional space will affect the sharpness of the telescope, is a small step generally more effective than a big step? Once we find a configuration that is almost good enough, it seems wise to avoid drastic movements in the high-dimensional space. Instead, fine-tuning around the current position seems to be a good strategy. Fisher proposed his geometric model to address similar questions in a different context—the evolution of a population.

Evolution of a population is similar to the exploratory design optimization of a telescope. A phenotype from a combination of genes at the genotypic level is a point in a high-dimensional space, just as the configuration of a telescope is a point in a high-dimensional space of the physical properties of a set of lenses. In Fisher's geometric model, each attribute of an organism is represented by a dimension in a Cartesian coordinate system. The optimum, the best combination of attributes, is located at the origin of the coordinate system. According to Darwinian evolution, if an environmental change shifts the population away from its optimum, the population will go through mutations that are phenotypically random with respect to the needs of organisms. That means that the phenotypic effect of a mutation may move toward or away from the optimum.

According to Fisher, adaptation is characterized by the movement of a population toward a phenotype that best fits the present environment. The phenotypic size of a mutation is the magnitude of its corresponding vector, that is, the distance to the origin, or the optimum. Fisher showed that the probability, $P_a(x)$, that a random mutation of a given phenotypic size r is favorable falls rapidly with mutational size. Mutations that are very close to the optimum are known as infinitesimally small mutations. Infinitesimally small mutations have a 50% chance of being favorable, but that rate drops rapidly as mutation size increases.

Fisher concluded that mutations small with respect to the optimum are the genetic basis of adaptation. Motto Kimura later pointed out that in addition to be beneficial, mutations must do something else to contribute to adaptation. They must escape accidental loss. Mutations of large effect are more likely to escape such loss than infinitesimally small mutations. Kimura suggested that mutations of intermediate size are the most likely to contribute to adaptation. However, the conclusion was also challenged later on. Computer simulations showed that adaptation in Fisher's model involves a few mutations of relatively large phenotypic effect and many mutations of smaller effect.

One of the surprising findings is known as the cost of complexity. The complexity of a species depends on the number of attributes involved. Complex species have more species than a simple one. During adaptation, the fitness of a complex species changes slower than that of a simple species. In part, the distance traveled to the optimum by a beneficial mutation is smaller in a complex species. As Charles Darwin claimed, "natural selection can act only by taking advantage of slight successive variations; she can never take a leap, but must advance by the shortest and slowest steps."

5.2.3   The Holey Landscape

Sewall Wright's original adaptive landscape is a surface in a multidimensional space that represents the mean fitness of the population given a combination at a genotype level. A point on the surface represents the population resulted from a combination in the high-dimensional space. The landscape metaphor leads to some intuitive interpretations. For example, peaks and valleys in a landscape indicate various possible states of the evolution of a population. On the other hand, the landscape metaphor along with the principles of natural selection imposes constraints on the movement of a population across the landscape. For example, natural selection would prevent the population from moving into a valley. A population may be stuck in a local optima on the top of a small hill and never be able to reach the highest peak across a valley. Furthermore, since a population is represented as a single dot on the landscape, it is not clear how the process of splitting a population into two different species would pan out in this framework.

Given the constraints imposed by selection, how can a population in an adaptive landscape move from one peak to another peak across a valley in between? Wright's own solution to this problem is his shifting balance theory, in which a small population may make random moves through genetic drifts to the foot of a fitness mountain and selection will make the population climb up the mountain from there. Sergey Gavrilets was not satisfied with this solution. He argued that the emphases on adaptive peaks and valleys by Wright's metaphor are largely due to our biases and experiences in the three-dimensional space instead of a space with a much higher dimensionality. Realistically, the dimensionality of adaptive landscapes is much larger than three, and they have very different properties from our intuition. Gavrilets proposed the idea of a holey landscape in 1999 to address some of these problems.

Gavrilets explained how a holey landscape could avoid the need to cross valleys in an adaptive landscape. An individual organism is formed by a combination of genes, and these combinations of genes form a genotypic space. A population will be represented by a cloud of points. Different populations or species will be represented by different clouds. These clouds can change their size, shape, and location to reflect changes made by selection, mutation, recombination, or other mechanisms of evolution. A fitness value can be assigned to each genotype to form an adaptive landscape. In the metaphor of a holey landscape, clusters of well-fit genotypes are highlighted, whereas all other genotypes are treated as holes in the landscape. The metaphor of a holey landscape allows the cloud of a population to transform into smaller clouds of subpopulations, which represents speciation. The reasoning is that selection will move the population away from the holes much faster than speciation. This metaphor would avoid the need to move across valleys of low-fit genotypes. The holey adaptive landscape is built on the assumption that combinations of genes tend to form clusters that extend throughout multidimensional genotype space and that these genotypes are not mutually compatible in that they are separated by "holes."

5.2.4 Kauffman's NK Model

A simple but powerful model of adaptation was proposed by Stuart Kauffman in 1993 at the sequence level of DNA. The model is known as the NK model, which is defined with two mathematical parameters N and K to characterize adaptive landscapes of a variety of ruggedness. The fitness surface is defined on the basis of all possible DNA sequences at a gene. Sequences are arranged in a way such that similar sequences are placed near to each other and substantially different ones are farther apart. The fitness of each sequence is the height of the adaptive landscape.

Although the NK models originated from evolutionary genetics, NK models are often adopted with the use of rules that are beyond biology, that is, rules that are not compatible with the principle of natural selection. For example, NK models use either random selection or gradient selection to find a beneficial sequence. NK models have received considerable attention in computer science and in the analysis of patents.

Suppose an entity is characterized by N features or attributes. Each attribute is a variable that can only have two possible values. The corresponding fitness landscape therefore contains a maximum of 2^N configurations. The variable K specifies how many other attributes will influence the overall fitness if one attribute is changing. The variable K characterizes the topology of the fitness landscape in two important ways:

- The value of K determines the smoothness or how rugged the fitness landscape is.
- The value of K influences the number of peaks in the fitness landscape. The number of peaks tends to increase with K.

The value of K characterizes the degree of epistatic interaction, which determines the smoothness of the fitness landscape. It specifies the extent to which other loci affect the fitness contribution of a given locus. If K equals 0, each attribute contributes to the overall fitness value, independent of other attributes. Given an attribute X, a small change in X corresponds to two adjacent values of X. The difference between fitness$(\ldots, x_i, \ldots)$ and fitness$(\ldots, x_{i+1}, \ldots)$ is bounded by $1/N$, which is the maximal fitness contribution of a single attribute.

On the other hand, if K increases, the landscape tends to lose its smoothness and become more rugged because a change in one attribute is now connected to the change of K different attributes. If $K = N-1$, then it corresponds to another extreme.

The fitness contribution of a single attribute depends on all the remaining $N-1$ attributes. Generally speaking, the maximal difference in fitness value between two sets of such attributes is $(K+1)/N$. In an NK model, the fitness landscape becomes increasingly rugged as the interaction among the underlying attributes increases (as the K increases). As a result, the number of local optima also increases.

The notion of a fitness landscape provides an intuitive basis of some of the most influential ideas from other areas. The original unit of analysis is biological organism. Subsequent applications of fitness landscape thinking have expanded to a much broader scope, ranging from individuals to organizations. For example, March and Simon's concept of neighborhood search was proposed in 1958. In terms of a fitness landscape, the search process is equivalent to a local hill-climbing effort if the height of the hill represents the fitness value of an ultimate target.

5.2.5 Local Search and Adaptation

In mathematics, a topology space is a mapping between a space X and a space Y that preserves the neighboring relationship. If points x_i and x_j in X are neighbors, they will remain to be neighbors through a topological mapping T in Y, that is, $T(x_i)$ and $T(x_j)$ are neighbors in Y. Topological properties of a space have implications on what one may infer relationships in X based on their relations in Y.

If a fitness landscape represents the fitness of organizations with different internal structures, the fitness level of a smooth landscape at a point associated with an organizational structure x is similar to the fitness level of a nearby point that corresponds to a similar organizational structure $x+dx$. In other words, a smooth fitness landscape conveys the message that organizations with similar structures would be similar in terms of their fitness levels. Scholars in the literature also refer to such relationships as correlations between fitness and organizational forms.

Since one would expect to achieve considerable changes in fitness by restructuring an organization, a local search in a rugged fitness landscape would make more sense than in a smooth one. A rugged fitness landscape implies that by making relatively small changes in the form of an organization, one may bring about substantial changes in terms of the overall fitness.

As early as the 1960s and 1970s, researchers suggested that adaptability is enhanced if there is a modest degree of interaction among the components of a complex system. For example, if the interaction among components of a system is too high or *tightly coupled*, a perturbation in one component would affect all other components in the system, and adaptation in such conditions would be improbable. If a critical component is fully dependent on other components, it will not be able to tolerate the failure of the components that it relies on. An intuitive model of the failure of a complex system is called the Swiss Cheese Model. Each component in the system is like a slice of Swiss cheese with a number of holes in it. Each hole is a potential hazard. The system as a whole can work perfectly fine, so long as none of the holes align through all the slices (Figure 5.2).

Adaptation is making internal changes so that a population fits the environment better. In contrast, dispersal is a strategy to increase fitness in a heterogeneous

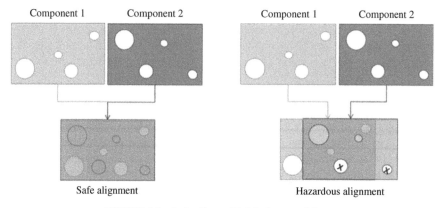

FIGURE 5.2   Swiss Cheese Model of system failures.

landscape by changing the environment in which an organism lives. Different habitat patches in the environment vary in terms of their expected fitness. The process of dispersal can be divided into three stages:

1. Emigration—the decision to leave the current habitat patch
2. Interpatch movement—the movement between patches
3. Immigration—the decision to enter and remain in a new patch

Optimal foraging theory and other theories have been proposed to explain factors that may affect decisions at various stages. For example, foraging individuals may want to minimize their energy consumption. One may attempt to optimize the amount of time they will spend in the current patch with reference to the quality of available food or other types of resources. Research that simulated the foraging behavior of monkeys suggested that they often prefer long trips to other high-quality trees over shorter trips to lower-quality trees. Each movement to a different patch seems to incur a substantial amount of overhead.

5.2.6   Criticisms

Wright's adaptive landscape diagram has been widely regarded as the most influential visual heuristics in evolutionary biology. Provine (Wright's biographer), however, argued in 1986 that the heuristics is poor because it is not mathematically interpretable.

Robert Skipper, a philosopher of biology at the University of Cincinnati, studied the work of Sewall Wright and came to the conclusion that when scientists face competing theories, their decisions are influenced not only by data and hypotheses, but also by additional factors such as fruitfulness and extensibility. Skipper considered earlier criticism of Wright's adaptive landscape metaphor, in particular, Provine's criticism that Sewall Wright's adaptive landscape heuristic has deep flaws and Rue's defense that Provine had not shown how it was flawed. Skipper pointed out that both Provine's and Rue's arguments were defective in light of his observation.

As we shall see shortly, the impact of Wright's adaptive landscapes has gone beyond evolutionary genetics and even biology as a whole. Adaptive landscapes are inspirational. The notion of fitness is extensible to be adopted for applications in a wide variety of distinct contexts. More importantly, especially from our perspective in this book, adaptive landscapes lend us a particularly suitable and generic framework for the study of gaps and how to bridge gaps through an optimization process.

5.3 APPLICATIONS OF FITNESS LANDSCAPES

We include three examples from different disciplines to illustrate the inspirational power of the fitness landscape framework. In the first example, the fitness landscape represents potency values of active compounds. In the second example, the fitness landscape represents the performance of companies. In the third example, the fitness landscape represents the impact of scholarly publications in terms of their citation frequencies.

5.3.1 Structure–Activity Relationship Landscapes

Structure–activity relationship (SAR) analysis is an important method in medicinal chemistry and pharmaceutical research. Structurally diverse molecules may have similar biological activities, whereas small changes of molecules often lead to dramatic changes in biological responses. If biological activities are seen as a function of the underlying molecular structure, then the activity surface may not be continuous; rather, it can be very rugged.

Preeti Iyer, Ye Hu, and Jürgen Bajorath published an innovative SAR analysis in 2011 in the *Journal of Chemical Information and Modeling*. Their SAR analysis is a novel application of activity landscape modeling. They monitored how compound composition, structural diversity, and global and local SAR features changed over time. Their study demonstrated how SAR characteristics of several datasets of compounds changed over time in terms of structure–activity landscape models. In particular, the results of their study revealed that compounds in their datasets have predominantly added around seed clusters of active compounds emerged early on, whereas other SAR islands remained largely unexplored. It seems to suggest that it is much harder than recognized to make long jumps in exploratory search on SAR landscapes.

A key element in the construction of their SAR landscape is a measurement called the SAR Index (SARI). It was proposed in 2007 by researchers from the same group, namely, Lisa Peltason and Jürgen Bajorath, to estimate how likely it is to identify structurally distinct molecules having similar chemical and biological activity. Using SARI, SAR relationships can be grouped into distinct categories in terms of continuity and heterogeneity. SARI characterizes SAR as continuous, discontinuous, heterogeneous relaxed, and heterogeneous constrained.

Given a set of active compounds and their potency values, SARI quantifies the systematic correlation of two-dimensional structural similarity and compound potency. It is calculated based on two scores: a continuity score and a discontinuity score. The continuity score measures the continuity of potency across structurally diverse compounds of high potency in terms of the potency-weighted mean of pairwise compound similarity. The similarity of the two compounds is calculated based on Tanimoto similarity for Molecular ACCess System (MACCS) fingerprint representations. The discontinuity score measures structurally similar compounds that differ considerably in their potency values. Thus, the discontinuity score identifies activity cliffs where small changes of structure may lead to large changes on the potency surface. Here the fitness of a compound is defined in terms of its potency in the context of pharmaceutical interest. More specifically, the discontinuity score is only calculated for compounds that have a Tanimoto similarity of 0.65 or higher and have at least one order or magnitude difference in potency.

The SARI of a set of compounds is defined as half of the sum of the normalized continuity score and one minus the normalized discontinuity score. For a predominantly continued SAR, its continuity score is close to 1, whereas its discontinuity score is close to 0. Therefore, the overall SARI is approximately (continuity + (1 − discontinuity))/2 ≈ (1 + (1 − 0))/2 = 1. The SARI of an essentially discontinuous SAR is approximately (continuity + (1 − discontinuity))/2 ≈ (0 + (1 − 1))/2 = 0. The SARI of a heterogeneous SAR, in which the continuity score and the discontinuity score are about the same approximately 0.5, is approximately (continuity + (1 − discontinuity))/2 ≈ (0.5 + (1 − 0.5))/2 = 0.5.

The activity landscape study used publicly available data from the ChEMBL database. The SAR analysis tracked the evolution of a few compound datasets, namely, adenosine A2a receptor (AA2), LCK tyrosine kinase (LCK), and μ-opioid receptor (MOR). Adenosine receptors, for example, include a few subtypes. Among them, the A2a receptor is believed to be related to the role of adenosine as an anti-inflammatory agent. Research is aimed toward the potential therapeutic role for A2a antagonists in treating Parkinson's disease. The AA2 dataset evolves in the study from the 53 compounds in 2000 to the 596 compounds in 2007. Corresponding SARIs increased from 0.03 to 0.56.

A network of molecules can be constructed based on Tanimoto similarity values. The nodes of two molecules are connected in the network if their Tanimoto similarity exceeds 0.80. Nodes can be color coded to represent corresponding potency, for example, from the lowest potency in the dataset in green to the highest potency in the set via yellow to red. The authors also chose to use the size of a node to represent the discontinuity score of the underlying compound. Thus, large nodes indicate that their potency differs from their structural neighboring compounds significantly. A group of large red and green nodes would reveal activity cliffs because of the discontinuity score and a mix of high and low potency levels.

The authors also generated three-dimensional activity landscape models, which consist of a base map of compounds and a fitness surface of their potency values. In essence, the fitness landscape represents the potency values of compounds. The distance between compounds in a dataset can be measured as the distance between two points in a high-dimensional space based on their structural similarities such as their MACCS fingerprints. A high-dimensional space can be projected to a lower-dimensional space with techniques such as multidimensional scaling. In the SAR analysis, the authors used nonmetric multidimensional scaling to project the interrelationships of a set of compounds to a two-dimensional base map. The potency values of these compounds are represented along the third dimension as the altitude of the resultant SAR landscape.

Figure 5.3 illustrates such SAR landscapes and how an SAR landscape changes as new compounds added to the dataset over time. The landscape in this example was generated based on the growth of the AA2 dataset. The white spaces seen in 2001 and 2003 essentially disappeared in 2005 and 2007 as new compounds were added to the A2a dataset. Green areas in the activity landscape are components with low potency, whereas red areas are compounds with high potency. The corresponding network in the same year provides additional information. For example, in 2001, the size of the red nodes indicates that there are sudden changes in potency values as minor variations in the structure were made. An important observation was that the further growth of the dataset was largely centered on four seed areas as shown in the figure. New compounds added to region 4 introduced a few major activity cliffs. This region becomes more promising for further exploration because it is more likely to find structurally similar compounds in the vicinity that may have a much higher potency yield. Figure 5.4 shows an enlarged view of the SAR landscape in 2005.

5.3.2   Landscapes Beyond Evolutionary Biology

In analogy to the role of a population in population genetics, the performance of a company can be characterized by multivariate measures. The fitness of a company can be captured by its intangible assets, such as its business strategies. Fitness can also be measured by operations, such as customer relationships and skills of employees. Thus, the fitness of a company at a particular time can be represented as a point on a fitness landscape. Its performance over an extended period of time can be intuitively shown by the movement of the trajectory of the company across the landscape.

In applications that have moved well beyond the territory of biology, the validity of evolutionary mechanisms originated in biology may need to be examined closely. We can no longer take for granted what role, if any, might be played by natural selection, mutation, and other evolutionary mechanisms in a new and fundamentally different context. On the other hand, as a mental model, adaptive landscapes offer an inspirational framework for developing a sound awareness of a complex situation, identifying an otherwise invisible gap between where we are and where we want to be, and making realistic plans to bridge the gap.

Intellectual Capital Landscapes

An intriguing example of how the idea of a fitness landscape can be extended beyond the scope of evolutionary biology is the work by Kitts, Edvinsson, and Beding in 2001. Their study appeared in *Expert Systems with Applications*. The construction of their novel landscape was based on how a company's performance can be measured and what would be the fitness value of a company.

The simplest performance measure of a company is probably its overall profit. The simplicity of the measurement, however, is less likely to reveal important information that is hidden by the complexity of how a company operates. In the early

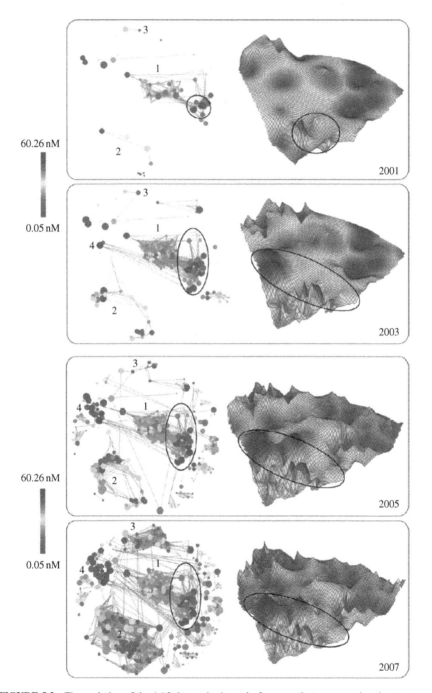

FIGURE 5.3 The evolution of the AA2 dataset is shown in four snapshots representing the structure–activity landscape in 2001, 2003, 2005, and 2007, respectively. Source: Reprinted with permission from Figure 3a in Iyer, Hu, and Bajorath (2011). © (2011) American Chemical society. (*See insert for color representation of the figure.*)

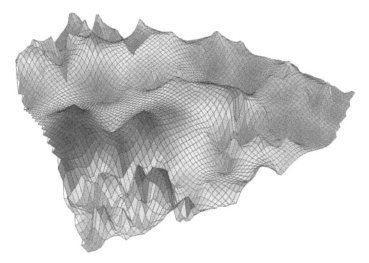

FIGURE 5.4 A close-up view of three-dimensional SAR landscape of the AA2 dataset in 2005. Source: Reprinted with permission from Figure S1 2005 in Iyer et al. (2011). © (2011) American Chemical Society. http://pubs.acs.org/doi/suppl/10.1021/ci100505m/suppl_file/ci100505m_si_001.pdf. (*See insert for color representation of the figure.*)

and mid-1990s, Leif Edvinsson, one of the coauthors of the 2001 study, proposed that the performance of a company should be measured from multiple perspectives, as diverse as possible. In particular, he devised a measurement called intellectual capital (IC), which measures the performance of a company in five distinct areas: financial, human, customer, process, and renewal. In the financial area, the company's performance is measured by the balance sheets. In terms of the human area, the skills and experience of the employees are taken into account. In the customer area, the relationships with customers of the company are evaluated and quantified. In the process area, the efficiency of internal functions contributes toward the overall performance evaluation. Finally, in the renewal area, the growth and long-term research and development potentials are estimated.

Kitts et al. constructed the fitness landscape of a group of companies in the following steps:

1. Using multidimensional scaling to generate a two-dimensional configuration of high-dimensional data about companies
2. Using fitness as the third dimension
3. Inferring the topographic surface around known points by interpolating techniques

The fitness landscape of a company represents a company as a single point. At each point on the fitness surface, the height represents the fitness of the state. A company's current state and its past states become visible on the landscape. One can study the trajectories of multiple companies such as their rivals or peers. Important areas on the landscape can be labeled clearly to serve as landmarks and guide the interpretations of various patterns with respect to these landmarks. For example, if it is known that a company performed exceptionally well or bad in a particular year, the position of the company provides a good reference point on the fitness landscape. Other companies may evaluate the trajectories of their own performance in terms of whether they are moving toward or moving away from the exemplar company.

Kitts et al. selected four companies from the large multinational Skandia investment group to study their movements on a generalized fitness landscape. The fitness, or state, of each of the companies was represented by its IC, which is an array of five variables that characterize the state in terms of financial, human skills, and other intangible attributes.

Kitts et al. developed a method to estimate the time taken for a company to move from one point on the landscape to another. To estimate the time, the calculation relies on the rate that a company's attributes changed historically. Suppose we are interested in the movement between two points, *a* and *b*, on the landscape. The estimated speed of the movement is the length of the path connecting *a* and *b* divided by the average change of all the attributes over time. The change of attributes is calculated in terms of absolute values.

The Skandia landscape map is shown in Figure 5.5 and Figure 5.6. The fitness surface represents the IC values. American Skandia moved across the landscape as it increased its fitness. Link, another company, remains to be stable in terms of its fitness. The performance of UK Life slipped down the slope on the landscape, which means it is losing its IC, that is, its fitness is reduced. Furthermore, it is possible to examine which specific aspects of the company are related to the decreasing fitness overall. The loss of IC was observed in 1997 with respect to its state in 1993 in three of the five major categories (i.e., parameters) in financial, customer, and renewal. In terms of standardized measures, the value of the customer category dropped from 1.19 to 0.27 from 1993 to 1997. Similarly, the value of renewal in standardized terms dropped from −0.39 to −0.98 in the same time period. The value of the financial fitness dropped from 1.64 to 0.41 during these years.

A Life Expectancy Landscape
Kits et al.'s study also demonstrated the flexibility of their approach with an example of a fitness landscape in a different context—the life expectancies of 44 countries.

The World Bank Organization (WBO) publishes the World Development Indicators Report each year. The WBO makes a wide variety of data available online. In particular, a broad range of country-level indicators are available on global development, poverty, the quality of life, environment, economy, finance, trade, and migration. For example, the economy table[3] includes a country's population, surface area, population density, gross national income, and gross domestic product. The 2013 edition was released on April 18, 2013.[4]

[3]http://wdi.worldbank.org/table/1.1
[4]http://databank.worldbank.org/data/download/WDI-2013-ebook.pdf

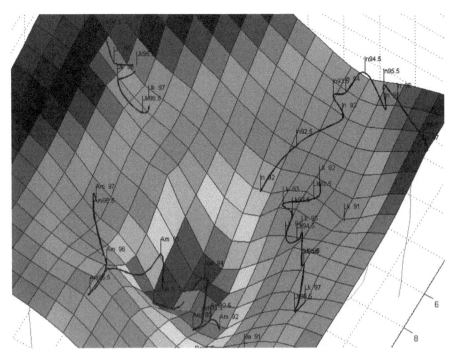

FIGURE 5.5 Skandia's IC landscape. Am, American Skandia; Di, Dial; In, Intercaser; UK, UK Life. Source: Reprinted from Kitts et al. (2001). © (2001), with permission from Elsevier.

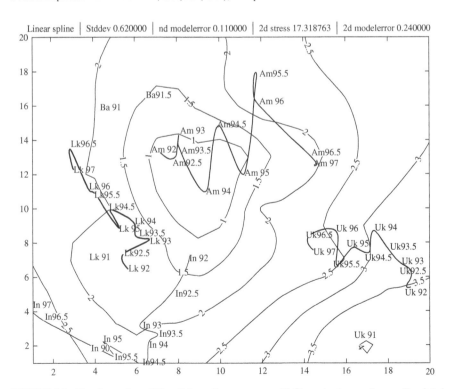

FIGURE 5.6 Top-down view of Skandia's performance on an IC-fitness landscape. Source: Reprinted from Kitts et al. (2001). © (2001), with permission from Elsevier.

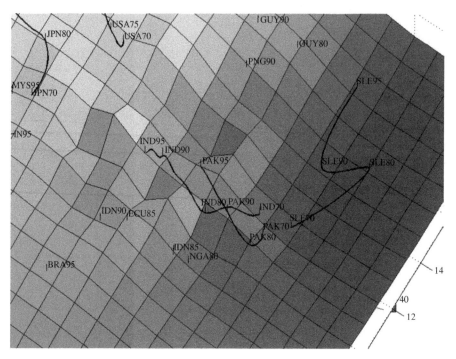

FIGURE 5.7 Fitness landscape of the life expectancy at birth in 44 countries from 1970 to 1995. Lighter colors represent longer life expectancy. Source: Reprinted from Kitts et al. (2001). © (2001), with permission from Elsevier.

In 2001, Kitts et al. constructed a fitness landscape of 44 countries based on 64 variables from the World Bank's 1998 report between 1970 and 1995 (see Figure 5.7). The fitness of a country was measured as the life expectancy of the country. Lighter colors represent longer life expectancy.

The life expectancy landscape of 44 countries reveals that many developing countries improved their life expectancy and moved to higher levels in the landscape. For example, India improved their life expectancies from 49 to 63 between 1970 and 1995. India's life expectancy trajectory started from the center of the landscape with its life expectancy in 1970 (IND70) and moved toward northwest. Its trajectory ended at its position in 1995 in a grid with a lighter shade of gray than the grid where IND70 was.

The authors commented on several issues that need to be addressed further. A significant issue is how to deal with the change of the landscape itself over time. In a retrospective study, the shape of the landscape is stable because it is based on hindsight with all the information available. In contrast, in a prospective study, the shape of the landscape would be subject to change as new information arrives. A world life expectancy landscape in 2013 is probably much different from the landscape in 1998. The trajectories on landscapes of different topography may become incomparable. If the fitness landscape represents our current belief, then a dynamic landscape is probably the norm as our belief is updated by new information we receive, for example, as we have seen in the example of a Bayesian search.

Unlike those fitness landscapes developed in the context of evolutionary genetics, the types of landscapes in Kitts et al.'s study did not come with additional constraints. For example, there is no constraining mechanism that functions like natural selection in evolution. A company may move freely on the IC landscape, and a country may travel anywhere on the life expectancy landscape. More fundamentally, the proximity of two positions in the low-dimension characteristic space is calculated on the surface value rather than supported by underlying mechanisms of change such as mutations. As the granularity rises from genotypic to phenotypic to molecular levels, interpretations of resultant landscapes depart from Wright's original paradigm. Adaptations of fitness landscapes beyond population genetics and evolutionary biology retain fewer and fewer of biological constraints.

Thematic Landscapes

Bovine spongiform encephalopathy (BSE) was first found in 1986 in England. A spongelike malformation was found in the brain tissue of affected cattle. The BSE epidemic in Britain reached its peak in 1992 and has since steadily declined. Creutzfeldt–Jakob disease (CJD) is an illness usually found in people aged over 55. It was first identified by two German neurologists in 1920 and has no known cause. Patients die about six months after diagnosis. New variant CJD (vCJD) is an unrecognized variant of CJD discovered by the National CJD Surveillance Unit in Edinburgh. vCJD is characterized clinically by a progressive neuropsychiatric disorder. Neuropathology shows marked spongiform change throughout the brain. The media reported a growing concern in the general public that BSE may have passed from cattle to humans.

While no definitive link between prion disease in cattle and vCJD in humans has been proven, the conditions are so similar that most scientists are convinced that infection by a BSE prion leads to vCJD in humans. The emergence of vCJD came after the biggest ever epidemic of BSE in cattle. The fact that the epidemic was in the United Kingdom and that most vCJD victims lived in Britain added to the evidence of a link. The British government assured the public that the beef was safe, but in 1996, it announced that there was possibly a link between BSE and CJD.

We conducted a search in literature to find out what scientific literature can tell us about the BSE, vCJD, and anything more. First, we generated a thematic landscape of the topic of BSE and CJD by searching the Web of Science with the term "BSE or CJD." Figure 5.8 shows a visualization of the search results in the form of a thematic landscape. The basis of the landscape is a network of cocited references. Two references are connected if they are cocited, that is, cited together by an article published later on. The fitness of a scientific publication represents its impact, which is often measured by the number of times subsequently published articles cite it. An article with a larger number of citations than others in the same field is usually considered to have a greater impact.

The visualization shown in Figure 5.8 is a snapshot of an animated landscape view. Each sphere represents a reference in the literature of a search topic. In this

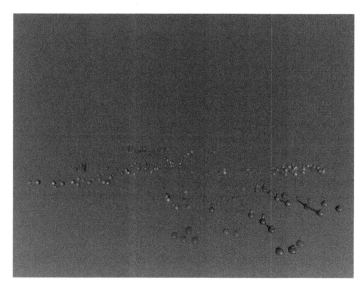

FIGURE 5.8 A rising landscape of research about mad cow disease (green), CJD in human (blue), and the fundamental research on prions (red). (*See insert for color representation of the figure.*)

case, the search query was "BSE and CJD." BSE is commonly known as mad cow disease. CJD is a similar disease in humans. The degree of transparency of a sphere represents the state of the underlying article at the time of a snapshot. Before the publication of an article, a search topic is represented by a translucent sphere. After its publication, the translucent sphere will become opaque. If the publication receives citations, then a vertical bar will grow from the sphere. The height of the bar represents the number of citations received up to the current time in the chronological animation of the history. The third dimension therefore represents the fitness of an individual article at a specific point of time. The animation also replicates the exact time when a cocitation link was added for the first time.

The landscape view in Figure 5.8 reveals three prominent areas of research, namely, mad cow disease (BSE) in the green area, CJD in the blue area, and prions. Prions are represented by the larger area in red and are believed to be the common cause of BSE and CJD. The area with a concentration of vCJD-related references has a very light color, which means that the area belongs to multiple areas of research.

Although our initial search was limited to BSE and CJD because we had never heard of anything about prions, the landscape of the resultant literature prominently revealed the role of prions in relation to the two topics. The concept of a prion—an infectious protein—was first introduced by Stanley Prusiner in a *Science* article (Prusiner, 1982). He was awarded a Nobel Prize in 1997 for his discovery of prions. Prions are an abnormal form of a protein, responsible for diseases such as scrapie in sheep, BSE in cattle—mad cow disease—and CJD in humans. These diseases are known as transmissible spongiform encephalopathy (TSE).

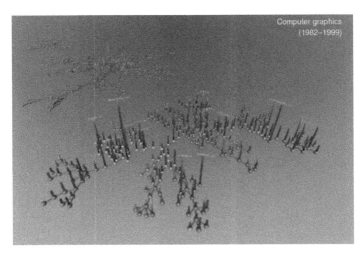

FIGURE 5.9 Thematic landscapes of computer graphics (1982–1999). (*See insert for color representation of the figure.*)

Figure 5.9 illustrates a composite thematic landscape of multiple fields of study in computer graphics. The fitness dimension represents the impact of scientific publications in terms of the number of citations they received. A peak in such landscapes represents a core area of research. The spectrum of colors on a citation bar indicates the distribution of citations over the period of time in which observations were made. An article with evenly distributed colors means that it has been receiving citations evenly in the past. An area of articles with a wide band of colors of recent years indicates an active and high-profile research topic.

Thematic landscapes in Figure 5.8 and Figure 5.9 provide a framework consistent with a dynamic fitness landscape. We will turn to the topic of predictive analytics shortly and address questions regarding things such as how to detect newly available information that can potentially transform the current landscape and how to track the movement of a population on thematic landscapes.

5.4 SUMMARY

We started this chapter with the legibility of cognitive maps. The primary role of a cognitive map is to help us navigate an environment and find what we need. Theories of optimal foraging and optimal information foraging underline the principles we tend to follow when we make decisions in a search. The fitness landscape metaphor and its variations provide a framework for analyzing where we are in the current situation and how we may reach where we want to be. Trajectories of populations, companies, and countries on a variety of fitness landscapes underline the fundamental value of the fitness landscape paradigm. It is the movement of a population, an organization, an idea, or an invisible college that we have the potential to adapt our current plans and optimize our actions in a complex adaptive system.

BIBLIOGRAPHY

Darken, R. P., Allard, T., & Achille, L. B. (1998). Spatial orientation and wayfinding in large-scale virtual spaces: An introduction. *Presence*, *7*(2), 101–107.

Dietrich, M. R., & Skipper, R. A., Jr. (2012). A shifting terrain: A brief history of the adaptive landscape (Chapter 1). In: S. Erik & C. Ryan (2012). *The adaptive landscape in evolutionary biology* (pp. 3–15). Oxford, UK: Oxford University Press.

Edvinsson, L., & Malone, M. (1997). *Intellectual capital: Realizing your company's true value by finding its hidden brainpower*. New York: HarperCollins Publishers, Inc.

Fleming, L., & Sorenson, O. (2001). Technology as a complex adaptive system: Evidence from patent data. *Research Policy*, *30*, 1019–1039.

Gavrilets, S. (1999). A dynamical theory of speciation on holey adaptive landscapes. *The American Naturalist*, *154*(1), 1–22.

Gavrilets, S. (2004). *Fitness landscapes and the origin of species* (Mpb-41). Princeton, NJ: Princeton University Press.

Gillespie, J. (1984). Molecular evolution over the mutational landscape. *Evolution*, *38*, 1116–1129.

Griliches, Z. (1990). Patent statistics as economic indicators: A survey. *Journal of Economic Literature*, *28*(4), 1661–1707.

Ingram, R., & Benford, S. (1995, October). *Legibility enhancement for information visualisation*. Paper presented at the 6th Annual IEEE Computer Society Conference on Visualization, Atlanta, GA.

Iyer, P. Hu, Y., & Bajorath, J. (2011). SAR monitory of evolving compound datasets using activity landscape. *Journal of Chemical Information and Modeling*, *51*, 532–540

Kaplan, J. (2008). The end of the Adaptive Landscape metaphor? *Biology & Philosophy*, *23*, 625–638.

Kauffman, S. A., & Levin, S. (1987). Towards a general theory of adaptive walks on rugged landscapes. *Journal of Theoretical Biology*, *128*, 11–45.

Kitts, B., Edvinsson, L., & Beding, T. (2001). Intellectual capital: From intangible assets to fitness landscapes. *Expert Systems with Applications*, *20*, 35–50.

Levine, M., Jankovic, I. N., & Palij, M. (1982). Principles of spatial problem solving. *Journal of Experimental Psychology: General*, *111*(2), 157–175.

Levine, M., Marchon, I., & Hanley, G. (1984). The placement and misplacement of You-Are-Here maps. *Environment and Behavior*, *16*(2), 139–157.

Levinthal, D. A. (1997). Adaptation on rugged landscapes. *Management Science*, *43*(7), 934–950.

Lynch, K. (1960). *The image of the city*. Cambridge, MA: The MIT Press.

Malan, K. M., & Engelbrecht, A. P. (2013). A survey of techniques for characterising fitness landscapes and some possible ways forward. *Information Sciences*, *214*, 148–163.

McCarthy, I. P. (2004). Manufacturing strategy: Understanding the fitness landscape. *International Journal of Operations & Production Management*, *24*(2), 124–150.

Orr, H. A. (2005). The genetic theory of adaptation: A brief history. *Nature Reviews—Genetics*, *6*, 119–127.

Palmer, A. C., & Kishony, R. (2013). Understanding, predicting and manipulating the genotypic evolution of antibiotic resistance. *Nature Reviews Genetics*, *14*, 243–248.

Peltason, L., & Bajorath, J. (2007). SAR index: Quantifying the nature of structure-activity relationship. *Journal of Medicinal Chemistry*, *50*(23), 5571–5578.

Provine, W. B. (1986). *Sewall Wright and evolutionary biology*. Chicago: University of Chicago Press.

Prusiner, S. B. (1982). Novel proteinaceous infectious particles cause scrapie. *Science*, *216*(4542), 136–144.

Scharnhorst, A. (2000). *Evolution in adaptive landscapes—Examples of science and technology development* (Discussion Paper FS II 00–302). Berlin, Germany: Wissenschaftszentrum Berlin für Sozialforschung.

Skipper, R. A., Jr. (2004). The heuristic role of Sewall Wright's 1932 adaptive landscape diagram. *Philosophy of Science*, *71*(5), 1176–1188.

Svensson, E., & Calsbeek, R. (2012). *The adaptive landscape in evolutionary biology*. Oxford, UK: Oxford University Press.

Thorndyke, P., & Hayes-Roth, B. (1982). Differences in spatial knowledge acquired from maps and navigation. *Cognitive Psychology*, *14*, 560–589.

Tolman, E. C. (1948). Cognitive maps in rats and men. *Psychological Review*, *55*, 189–208.

Wright, S. (1932). The roles of mutation, inbreeding, crossbreeding, and selection in evolution. In: *Proceedings of the sixth international congress on genetics*, Ithaca, NY (pp. 356–366).

Zahn, C. T. (1917). Graph-theoretical methods for detecting and describing Gestalt clusters. *IEEE Transactions on Computers, C, 20*, 68–86.

CHAPTER 6

Structural Variation

6.1 COMPLEX ADAPTIVE SYSTEMS

John H. Holland defines a complex adaptive system (CAS) as a system that has a large number of components that interact, adapt, or learn. These components are often called agents. Many phenomena in the world share essential properties of a CAS. The study of CAS focuses on complex, emergent, and macroscopic properties of such systems. A CAS typically involves a large population of agents that interact with each other. Other properties of a CAS include open and blurred boundaries, a constant flow of energy to maintain its organization, autonomous agents, and self-organizing mechanisms such as feedback.

6.1.1 Early Signs of Critical Transitions

Up to the seventeenth century, white swans were the only swans that had ever been seen in Europe. No one knew whether there was such thing as a black swan until they were found in Australia. Black swans, in Nassim Nicholas Taleb's *The Black Swan*, are highly improbable and unpredictable events that have a massive impact. The major myth of a black swan is that they are impossible to foresee.

About 34 million years ago, after being in a tropical state for many million years, the Earth suddenly changed to a colder state, a transition known as a greenhouse–icehouse transition. Epileptic seizures are a transient symptom often associated with a sudden contraction of a group of muscles. Sudden shifts from one state to another can be found in many complex dynamical systems. A key question is whether there are early warning signs before such sudden and radical changes occur.

Ecosystems, financial markets, and the climate are all complex dynamical systems. Asthma attacks and epileptic seizures are examples of spontaneous systemic failures. A sudden collapse of a civilization due to overpopulation is another type of example. These types of system-wide changes are often characterized as state transitions.

The Fitness of Information: Quantitative Assessments of Critical Evidence, First Edition. Chaomei Chen.
© 2014 John Wiley & Sons, Inc. Published 2014 by John Wiley & Sons, Inc.

It may be more likely for a system to change from one state to another. A global change may or may not be desirable. Some changes are not reversible. For instance, many are concerned that the current climate change might lead to a possibly irreversible catastrophic change of the entire ecosystem.

Interestingly, the predictability of a financial market is something that the financial system would like to eliminate rather than enhance. One example of an early warning signal is a measure of increased trade volatility. In contrast, scientific revolutions, as one would expect from high-risk and high-payoff transformative research, are intended and desirable.

In 2009, *Nature* published a review article on early warning signs for critical state transitions in complex dynamical systems (Scheffer et al., 2009). The review was written by 10 authors from four countries—Germany, the Netherlands, Spain, and the United States—and multiple disciplines such as environmental sciences, economics, oceanography, and climate impact research.

Critical transitions can be explained in terms of bifurcations. A bifurcation is when a small smooth parameter change causes a sudden qualitative change in a system's behavior. On the surface of a fitness landscape, a bifurcation will be shown as a cliff. A small change in the genotype may cause a big change in the phenotype. A small variation in the fingerprint of a compound may lead to a big jump in its potency. Adding a single link to a complex network may bridge conceptual gaps in a revolutionary way.

Predicting critical transitions in advance is extremely difficult because the system may show little change before the system suddenly shifts to a different state, just like how seeing white swans all the time will not tell us anything about the arrival of a black swan. Traveling on a smooth landscape may suddenly encounter a cliff.

A fundamental question is whether the system gives any early signs at all as it approaches to critical transitions. Scheffer et al.'s review in *Nature* indicated that although predicting critical transitions in advance is extremely difficult, research in several different scientific fields suggests that generic early warning signals may exist. If the dynamics of system near a critical point has generic properties regardless of differences in the details of each system, then it is indeed a profound finding.

The most important clues whether a system is getting close to a critical threshold are related to a phenomenon called *critical slowing down* in dynamical system theory. To understand critical slowing down, we need to explain a few concepts, such as fixed points, bifurcations, and fold bifurcations. A *fixed point*, also known as an invariant point of a function, is a point that is mapped to itself by the function. As far as the fixed point is concerned, it remains unaffected by the function or mapping. Fixed points are used to describe the stability of a system. In a dynamical system, a *bifurcation* represents the sudden appearance of a qualitatively different solution for a nonlinear system as some parameters change. A bifurcation is a separation of a structure into two branches.

At fold bifurcation points (e.g., one stable and one unstable fixed points), the system becomes increasingly slow in recovering from perturbations. Research has shown that (1) such slowing down typically starts far from the bifurcation point and (2) recovery rates decrease smoothly down to zero as the critical point is approached.

The change of the recovery rate provides important clues of how close a system is to a critical transition. In fact, the phenomenon of critical slowing down suggests three possible early warning signals in the dynamics of a system approaching a radical change: slower recovery from perturbations, increased autocorrelation, and increased variance.

Scheffer et al.'s review emphasizes that the work on early warning signals in simple models is quite strong and it is expected that similar signals may arise in highly complex systems. They also note that more work is needed, especially in areas such as detecting patterns in real data and dealing with challenges associated with handling false positive and false negative signals. It is also possible that sudden shifts in a system may not necessarily follow a gradual approach to a threshold.

6.1.2   Early Signs of Great Ideas

Detecting early signs of potentially valuable ideas has theoretical and practical implications. For instance, peer reviews of new manuscripts and new grant proposals are under a growing pressure of accountability for safeguarding the integrity of scientific knowledge and optimizing the allocation of limited resources (Chubin, 1994; Chubin & Hackett, 1990; Häyrynen, 2007; Hettich & Pazzani, 2006). Long-term strategic science and technology policies require visionary thinking and evidence-based foresights into the future (Cuhls, 2001; Martin, 2010; Miles, 2010). In foresight exercises on identifying future technology, experts' opinions were found to be overly optimistic on hindsight (Tichy, 2004). The increased specialization in today's scientific community makes it unrealistic to expect an expert to have a comprehensive body of knowledge concerning multiple key aspects of a subject matter, especially in interdisciplinary research areas.

The value, or perceived value, of an idea can be quantified in many ways. For example, the value of a good idea can be measured by the number of lives it has saved, the number of jobs it has created, or the amount of revenue it has generated. In the intellectual world, the value of a good idea can be measured by the number of other ideas it has inspired or the amount of attention it has drawn.

What patterns and properties of information can tell us something about the potential values of ideas expressed and embodied in scientific publications? A citation count of a scientific publication is the number of times other scientific publications have referenced the publication. Using citations to guide the search for relevant scientific ideas by way of association, known as citation indexing, was pioneered by Eugene Garfield in the 1950s (Garfield, 1955). A consensus has been reached that citation behavior can be motivated by both scientific and nonscientific reasons (Bornmann & Daniel, 2006). Citation counts have been used as an indicator of intellectual impact on subsequent research. There have been debates over the nature of citations and whether positive, negative, and self-citations should all be treated equally. Nevertheless, even a negative citation makes it clear that the referenced work cannot be simply ignored.

What do we know about factors that may influence citation counts in one way or another? An article that has been highly cited so far is likely to remain highly cited according to the Matthew Effect (Merton, 1968). An article that has been frequently downloaded or viewed online is likely to become highly cited later on (Brody & Harnad, 2005; Kurtz et al., 2005). Relying on direct evidence such as visit counts, download counts, and citation counts that an article has already obtained has relatively lower risks than making assessments based on indirect evidence. The focus of these approaches, however, is on a trend that is already observable. These methods will not be able to tell us anything about a newly published article, which has not been known to the scientific community long enough to be viewed, downloaded, or cited.

Researchers have searched for other signs that may inform us about the potential impact of a newly published scientific paper, especially those that can be readily extracted from routinely available information at the time of publication instead of waiting for download and citation patterns to build up over time. Factors such as the track record of authors, the prestige of authors' institutions, and the reputation of the journal of the article are among the most commonly considered ones (Boyack, Klavans, Ingwersen, & Larsen, 2005; van Dalen & Kenkens, 2005; Hirsch, 2007; Kostoff, 2007; Walters, 2006). The common assumption central to approaches in this category is that great researchers tend to continuously deliver great work and, along a similar vein, an article published in a high-impact journal is likely to be of high quality itself.

The sources of information used in these approaches, however, are not direct or specific to the ideas reported in a new publication. In an analogy, we totally rely on the credit history of an individual, but do nothing to assess the risk of the current transaction. Furthermore, with such approaches, we will not be able to know what precisely makes an idea novel. We will not be able to know whether similar ideas have been proposed in the past.

6.1.3   The Structural Variation Theory

Scientific knowledge is in constant interplay with new ideas proposed by the scientific community. A critical part of scientific inquiry is to discern where a new idea stands with reference to the collective knowledge of the scientific community as a whole. This is a cognitively demanding and conceptually challenging task because scientific publications appear faster than anyone could possibly read, analyze, and synthesize. Not only do we need to have an up-to-date and comprehensive understanding of the intellectual structure, but also to identify exactly how a newly proposed idea would fit to the existing structure. Figure 6.1 illustrates how the 2002 study by Galea et al. on the prevalence of posttraumatic stress disorder (PTSD) fits to the previous belief of the research community:

1. The domain knowledge prior to Galea et al.'s study believes that PTSD is caused by directly experiencing a trauma.
2. Galea et al.'s study states that indirect experience is found to cause PTSD, which is an unprecedented connection.
3. The new connection is incorporated into the domain knowledge.

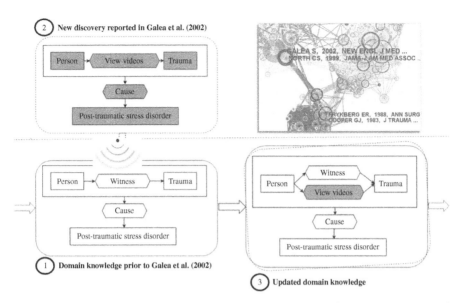

FIGURE 6.1   A newly found connection changes the existing structure of the domain knowledge of PTSD.

The complexity reduction process shown in Figure 6.1 is challenging to achieve. Many ideas are complex, and each component may involve varying degrees of ambiguity and uncertainty. Although using structured abstracts is likely to increase the clarity of communicating a scientific contribution effectively, the current conventions of scholarly communication may not encourage authors to cut to the chase and present their work in a concise and accurate manner. More fundamentally, a substantial degree of domain knowledge is required to discern the essence of a discovery and articulate it in easy-to-understand languages. Nevertheless, if the Nobel Committee can give clear and easy-to-understand reasons why Nobel Prizes are awarded to complex scientific contributions, would it be reasonable to expect the same level of clarity and simplicity when scientists contribute to not only the growth of the volume of knowledge, but also its clarity?

An additional benefit of reducing the complexity of scientific literature is that it may become easier for us to see where no man's land is, where the gaps are, and how they might be bridged.

The development of scientific knowledge is a process of interplay between the intellectual structure and a stream of incoming new ideas conveyed in newly published scientific papers. The structural variation theory aims to provide specific trails of evidence to show why and how an idea is novel with reference to the current intellectual structure of a scientific domain.

Each new idea is a potential perturbation to the system of existing ideas. As illustrated in Figure 6.2, the structure and content of a knowledge system form a CAS. An individual's mental model is a knowledge system. The collective beliefs of a community of scientists are also a knowledge system. Scientific knowledge as

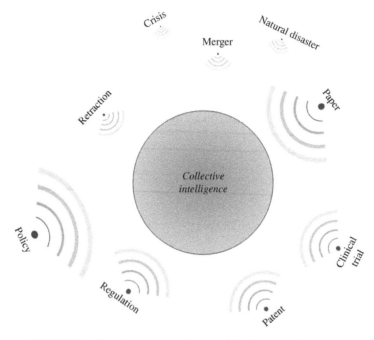

FIGURE 6.2   Signals may cause critical transitions in a CAS of knowledge.

a whole is a dynamic system because it is subject to the perturbations caused by a wide variety of new information, such as novel theories in newly published articles, new ideas in patent applications, surprising results in clinical trials, controversial new policies and regulations, and even newly announced retractions of previously published studies. These signals may cause a critical transition in the system, or they may not impact the system at all. It is also possible that they generate an impact that is not as extreme as the other two scenarios.

The potential value, or the impact, of an idea can be measured in terms of the extent to which the new idea changes the structure of a knowledge system. The significance of such changes can be further measured at a local level as well as at a global level. We refer to the overall structure-changing potential of a signal as its transformative potential. Figure 6.3 illustrates (a) incremental changes and (b) transformative changes or critical transitions in a CAS. Incremental changes are easier to predict than transformative changes. For instance, a highly cited article is likely to be continuously cited.

The structural variation theory characterizes the potential impact of structure-changing mechanisms based on whether a newly proposed link is unprecedented and whether the scope of impact is local or global. If a new idea connects previously disparate patches of knowledge, then its transformative potential is higher than the potential of ideas that are limited to well-trodden paths over the existing structure. The central idea in the structural variation theory is a boundary-spanning and gap-bridging mechanism.

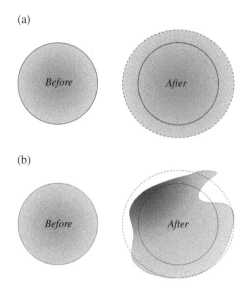

FIGURE 6.3   Incremental changes (a) and transformative changes (b) caused by a perturbation to the system. (*See insert for color representation of the figure.*)

A system may consist of components organized at various levels of granularity or aggregation. For example, scientific knowledge can be organized by disciplines at the top level, by fields within individual disciplines, by topics within a field of study, and by propositions within a topic area. The existing structure may be changed at each of these levels. From the system's point of view, a change at a higher level is likely to be fundamental to the overall behavior of the system than a change at a lower level would be. A new discovery that leads to the creation of a new discipline would have a much higher impact than an experiment that replicates the results of a famous experiment. Figure 6.4 illustrates how we differentiate the potential impact of new information at various levels of granularity. With reference to the existing structure, new information may bring changes in several ways, including adding a new node, removing an existing node, adding a new link, and removing an existing link. In the following descriptions, we will focus on estimating the impact of links because adding or removing nodes can be approximated by link changes. Adding an unprecedented link is denoted by a dashed line. The removal of an existing link is represented by a solid line with a cross. Suppose the system has three clusters of nodes at the top level. Link changes at the cluster level are weighted more than link changes at the individual node level. Specifically, event 1 is adding an unprecedented long-range link between clusters. It is the strongest signal to take into account when estimating its potential for changing the behavior of the system as a whole. Event 2 is removing an existing link between clusters. It is considered the second strongest signal. Event 3 is reinforcing an existing link at the cluster level, which sends the weakest signal in estimating its transformative potential. At the individual node level, there are three similar types of link change events in the order of decreasing signal

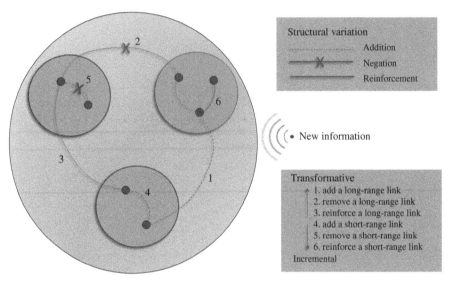

FIGURE 6.4 Boundary-spanning mechanisms modeled in the structural variation theory. (*See insert for color representation of the figure.*)

strengths, namely, event 4, adding an unprecedented link within a cluster; event 5, removing an existing link inside a cluster; and event 6, reinforcing an existing link inside a cluster.

In the context of scientific literature, nodes in the system of scientific knowledge represent references of publications, and links represent evidence of cocitation relations found in literature. A new article typically cites a subset of the references in the system. In effect, the new article sends a series of signals about cocitation links. If an article does not generate any new links, then its transformative potential is considered to be low. In contrast, if an article proposes many unprecedented connections among references, then it would have a high transformative potential, provided that the author of the article is not totally ignorant of existing knowledge and is not trying to contaminate the system.

One way to validate the structural variation theory is to use the estimated potential value of an article to predict the number of times it will be cited in the future. The accuracy of the prediction will inform us the importance of the structural changes caused by the new information. Furthermore, we can compare the predictive effects of structural variation linkage and the more traditional and less direct measures such as the number of coauthors and the number of references cited by the article.

The structural variation theory offers a conceptually simple and unifying explanation of several commonly identified citation predictors. For instance, review and survey articles are often highly cited because they are likely to synthesize multiple areas in a broader context than original research articles. As a result, they are more likely to create boundary-spanning connections. Similarly, one can infer from the theory that an article contributed by more coauthors is more likely to reflect a diverse range of perspectives and expertise than articles of fewer coauthors. This increased diversity in turn is more likely to create unprecedented links at a higher level of aggregation.

6.2 RADICAL PATENTS

A patent is a disclosure of technical inventions. In return, its intellectual properties are protected by the issuing government agency to exclude anyone else from using the patented content, which could be device, apparatus, or process. The patent system aims to encourage invention and technical progress by providing a temporary monopoly for the inventor and by forcing the early disclosure of the technical details. What is patentable? What can we learn from the disclosed information? These questions do not have straightforward answers. In fact, numerous disputes have to be settled legally at courts of the highest level, such as the U.S. Supreme Court.

6.2.1 Patentability

Bernard L. Bilski and Rand A. Warsaw filed a patent application of a procedure for protecting buyers and sellers from price fluctuations in economy. Their application was rejected by the U.S. Patent and Trademark Office (USPTO). The claims in dispute are claims 1 and 4. Claim 1 describes a series of instructions of how to hedge risk. Claim 4 expresses the concept in Claim 1 in a simple mathematical formula. The USPTO's patent examiner rejected the patent application on the ground that it is not implemented on a specific apparatus and it merely manipulates an abstract idea. Moreover, it solves a purely mathematical problem without using the technological arts.

They appealed to the Court of Appeals. The Court of Appeals ruled that the claims are not patentable because it failed the machine-or-transformation test, which was regarded as the sole test of patentability by the Court of Appeals. The test was described in the Patent Act, 35 U.S.C. Section 101. The case was then taken to the U.S. Supreme Court. The Supreme Court reached a decision on June 28, 2010, on the *Bilski v. Kappos* case, 130 S.Ct 3226, (2010).

The Supreme Court urged students of patent law to study these scholarly opinions. The Supreme Court ruled that the issue was not about whether or not the machine-or-transformation test was the exclusive test, but instead was that this was an attempt to patent abstract ideas.

According to the Supreme Court opinions, the precedent cases in the Supreme Court established that the machine-or-transformation test is a useful clue to determine whether claimed inventions are processes under Section 101, but is not the sole test for deciding the patentability of an invention. The test may be more suited for the industrial age, but it may not be adequate for the information age. In short, the Court of Appeals made the right decision but gave the wrong reason.

The Supreme Court declined to comment on the patentability of a particular invention. Rather, it explained the principles about how the old patent law should be interpreted so that the ultimate purpose of the patent system can be served properly in subsequently emerged unforeseeable cases. The greatest challenge is how to strike the balance between protecting inventors and not granting monopolies over obvious inventions, that is, procedures that could be reached by applying general principles.

An abstract idea, law of nature, or a mathematical formula is not patentable. However, an application of general knowledge from the list may be patentable. This principle was established by the Congress when it created the USPTO to encourage and reward inventions. Abstract ideas, law of nature, and mathematics are regarded as properties of the human race. It follows that no monopolies established on the knowledge would be justifiable because they would hurt inventions. Limiting an abstract idea to a particular domain of application does not make the concept patentable, which was essentially the nature of the *Bilski v. Kappos* case as seen by the Supreme Court.

The ruling on the *Bilski v. Kappos* case was influential on numerous cases that subsequently emerged. A few examples are as follows:

- *CyberSource Corp. v. Retail Decisions, Inc.*, 654 F. 3d 1366 (2011), Court of Appeals, Federal Circuit
- *Duarte-Ceri v. Holder*, 630 F. 3d 83 (2010), Court of Appeals, 2nd Circuit

The bottom line is that a patentable idea needs to be novel and useful. It cannot be an abstract idea, which means a generic idea that can be applied to many unforeseen areas.

6.2.2 NK Models of Recombinant Patents

Organizational theorists have extended the idea of fitness landscapes from population genetics to the study of organizations. We have introduced NK models in Chapter 5. Here we discuss some examples of how NK models are used in the context of patent analysis.

Kauffman's NK models of fitness landscapes provide an inspirational framework for the theory of invention. Just as evolution is conceptualized as a search over the fitness landscape in evolutionary biology, inventions can be similarly conceptualized as a search over a landscape shaped by various possible combinations of ideas. Since the efficiency of a search is a function of the topography of such landscapes, this framework offers a systematic way to study the evolution of inventions.

Lee Fleming and Olav Sorenson published a study in Research Policy in 2001. They conceptualized technological invention as the process of a recombinant search over technology landscape. In particular, they used Kaufman's NK models to develop a systematic and empirically validated theory of invention. They also noted that the process of invention differs from biological evolution in important ways.

The new metaphor of invention implies that the trajectory of search is continuous on the landscape. As the search path extends, new inventions are encountered and used. Is it possible that a trajectory could be discontinuous with respect to a given landscape? Fleming and Sorenson did not particularly address these issues in their study.

In an analogy to genetics, an invention's components are like an organism's genes. An invention is a combination of some components. Different ways to combine components will lead to different inventions. A new invention may emerge by reorganizing the components of existing inventions or by combining new and existing components.

It has long been considered that technological novelty is likely to come from the recombination and synthesis of existing ones, although there are certainly other forms of novelty. Using the automobile as an example, Fleming and Sorenson conceptualized it as a combination of the bicycle, the horse carriage, and the internal combustion engine.

To Fleming and Sorenson, the key insight from Kauffman's NK model is how interdependence increases the complexity of the fitness surface. As K increases, the fitness surface becomes increasingly rugged. In other words, with high interdependency, a small step in the underlying genotype space could lead to abrupt shifts of fitness. If this is also true in the invention analogy, one would expect that a small variation in how ideas are recombined may lead to inventions that may differ substantially in terms of their novelty and utility—the ultimate measure of the fitness of a technological invention.

The number of components to be combined by a new invention corresponds to the N in Kaufman's NK model, whereas the interdependence among these components corresponds to the K parameter. Kauffman's work suggested that an intermediate level of interdependence is optimal for adaptation. The probability of conflicting constraints increases as the interdependency increases. It becomes difficult to make overall improvements by changing only one dimension.

Fleming and Sorenson analyzed 17,264 U.S. patents granted in May and June 1990. The fitness or usefulness of these patents was measured in terms of citations to these patents made by other patents. An NK model was constructed with N as the number of subclasses assigned to each patent and K as the recombination for each patent's subclasses. Fleming and Sorenson defined the fitness of a patent as the number of citations the patent receives six years and five months after its grant date, mainly because of the consideration that citations to a patent typically peak three to five years after the grant date.

The value of N represents the number of subclasses assigned to a patent. The value of the interdependence K was calculated in two steps. First, the ease of recombination of a subclass i E_i as the number of subclasses previously combined with subclass i divided by the total number of patents in subclass i. The value of E_i will be larger for a subclass that has been widely used than a subclass that has been narrowly used if they cover the same number of patents. The second step calculates the interdependence of a patent j K_j as the number of subclasses on patent j divided by the sum of the E_i for subclasses in which patent j is assigned to.

Each patent cites existing patents in two categories: one category contains references to patents made by the patent examiner, whose primary goal is to establish the novelty of the current patent with respect to prior art patents; the other category contains references to patents and other publications made by the inventors themselves.

Fleming and Sorenson analyzed patent citations with negative binomial (NB) models. Their results showed that the key factors in the NK model only imply marginal practical importance. They interpreted the results as a divergence from the predictions of the NK model.

Their study found that local search affects the outcome of invention. More importantly, the interdependence and size of the search space impact the likelihood of successful search more than any other characteristics of the invention process. The effect of K and N together can make the difference between a median invention and one in the top 6% in terms of patent citations.

The results of their study also raised a more fundamental question about the extent to which technological evolution differs from biological evolution. In Kauffman's models, search patterns are simplistic, including hill climbing and random combination. In contrast, search heuristics in invention are expected to be much more complex. Fleming and Sorenson identified the need for a better understanding of the mechanisms that are involved in creative activities such as innovation and invention.

Fleming and Sorenson are convinced that intelligent agents such as inventors do not blindly recombine components. They suggested that learning from experience and other heuristics should be taken into account in further research along the adaptation of Kauffman's NK models.

6.2.3   Recombinant Search for High-Impact Radical Ideas

Although finding new insights is the ultimate goal of researchers in many fields, there has been a lack of a mechanism that one can adopt. The concept of an insight is widely used, probably with many different meanings. From a methodological point of view, the notion of an insight encapsulates too much information to be useful when we seek a concrete procedure that we can execute step by step. As Metcalfe noted in 1995, "We do not yet understand insight."

Melissa A. Schilling and Elad Green published an interesting article in 2011. The article shed light on what types of mechanisms might be involved in the generation of high-impact ideas. They studied three factors that may contribute to the generation of high-impact ideas in social sciences.

The premise of their study is that the current knowledge of a domain can be seen as a network of elements, and when new information is received, it may alter how the existing elements are connected in such a way that dramatic changes may emerge at the system level. Recombinant search provides an exploratory process of finding new ways to expand the network of knowledge and change the connectivity of the structure. Schilling and Green's premise shares the same principle as our structural variation theory. Bridging the right gap is the key!

There seems to be three ways to achieve the goal. First, if the search covers a broad range of potential candidates, then one may be exposed to a wide variety of interpretations and different perspectives, thus the search is more likely to generate novel ways to combine available information and form a new knowledge network. Second, the depth of the search characterizes a focused area of knowledge. Third, atypical connections are new links that are largely unexpected to appear. Atypical connections would appear to be odd.

A pragmatic insight from recombinant search is that atypical or unexpected combinations of knowledge can lead to novel outcomes with greater variance in performance than the "usual" and more expected ones. In other words, adapting an unexpected combination of knowledge may lead to insights that can radically change our understanding of a problem as much as mislead us to a less meaningful path. The role of atypical combinations in recombinant search has been justified in terms of potential advantages of accessing to alternative perspectives. A diverse range of possibly competing and heterogeneous interpretations may lead to useful reconceptualizations of problems. In terms of mental models, considering atypical linkage is critical to evaluate the current mental model, the mindset, or a paradigm. If a new atypical link can trigger the shift of such models, then it would be the best one can expect. We should also be mindful that many cases do not conform to the best-case scenario.

Schilling and Green's method is of particular interest because of its methodological novelty in its own right. They sampled breakthrough publications from Thomson Scientific's high-impact papers database in four disciplines of social sciences, namely, economics, management, psychology, and sociology. They did not seem to mention why they limited their study to social sciences, although their method appears to be applicable to a much broader range of disciplines.

From each of the four disciplines, they selected 10 papers with the highest impact. These 40 high-impact papers were then paired with a control group of 80 papers. Each high-impact paper was matched by two control papers published in the same journal in the same year. Schilling and Green called this random-but-matched pairing. The resultant 120-paper dataset was then analyzed for patterns associated with recombinant search in terms of the scope, the depth, and the atypical linkage.

What was novel and unique in their design is the use of the Dewey decimal system as the template to measure the scope and the depth of recombinant research and estimate whether a particular linkage would be rare or common. The Dewey decimal classification system was created by Melvil Dewey in 1876. It is probably the most widely used system for classifying library collections by topic. The structure has been revised to keep up with the modern needs. Currently, the Online Computer Library Center (OCLC) is responsible for the maintenance of the Dewey decimal system.

The Dewey decimal system is a three-level hierarchical classification scheme. The highest level has 10 classes to represent major disciplines. The next level is known as the Hundred Divisions. The top-level classes are divided into categories. The first digit of these subclasses shares the same first digit of their parent top class. For example, political science (320), economics (330), and law (340) all belong to the top-level class social sciences (300). Table 6.1 shows a few more examples.

The third level is known as the thousand sections. Here are some examples of the third-level sections under Mathematics (510), which in turn belongs to the top-level class of Science (500) (see Table 6.2).

Schilling and Green used the OCLC WorldCat system to find the corresponding Dewey decimal numbers of the references cited by the 120 sample papers. They were

Table 6.1   The first two levels of the Dewey decimal system[a]

Main classes		The hundred divisions	
000	Computer science, information, and general works	010	Bibliographies
		020	Library and information sciences
		030	Encyclopedias and books of facts
		030	[Unassigned]
		…	……
100	Philosophy and psychology	110	Metaphysics
		…	……
		150	Psychology
		…	……
200	Religion		
300	Social sciences	310	Statistics
		320	Political science
		330	Economics
		340	Law
		…	……
400	Language		
500	Science	510	Mathematics
		520	Astronomy
		530	Physics
		540	Chemistry
		550	Earth sciences and geology
		560	Fossils and prehistoric life
		570	Life sciences, biology
		580	Plants (botany)
		590	Animals (zoology)
600	Technology	610	Medicine and health
		620	Engineering
		630	Agriculture
		…	……
700	Arts and recreation	…	……
		780	Music
		790	Sports, games, and entertainment
800	Literature, rhetoric, and criticism		
900	History and geography		

[a] Source: From OCLC. http://www.oclc.org/dewey/resources/summaries.en.html

able to match 5318 cited references, which is 79% of all the references cited by the sample papers.

The dependent variable represents whether a paper was from the high-impact group or from the control group. The independent variables include the depth of recombinant search, the scope of the search, and atypical combinations.

Table 6.2   The third level sections under science (500) in the Dewey classification system

511	General principles of mathematics
512	Algebra
513	Arithmetic
514	Topology
515	Analysis
516	Geometry
517	[Unassigned]
518	Numerical analysis
519	Probabilities and applied mathematics

Since each paper cites an array of references, the concentration of the references cited by the paper can be measured by the Herfindahl–Hirschman Index (HHI). The HHI measure is the sum of the squared percentage of references c_i from each of the second-level Dewey classification divisions:

$$\text{HHI(paper)} = \sum_i c_i^2 \quad \text{where} \quad \sum_i c_i = 1$$

For example, if a paper cites references equally from four Dewey divisions, then HHI $= 0.25^2 \times 4 = 0.25$. If a paper cites all references from a single division, then its HHI is 1, which is also the maximum possible value of HHI. If it cites references evenly from n divisions, then the percentage from each division is $(1/n)$. The HHI is $(1/n)^2 \times n = (1/n)$. Thus, the HHI measures the concentration of the citation with respect to the divisions in the Dewey system. The depth of a paper's recombinant search is defined as the HHI times the number of references (NR) in the paper and logarithmically transformed to improve its normality: $d = \log(\text{HHI} \times \text{NR})$.

Schilling and Green defined the scope of search by assigning points to Dewey classes, divisions, and sections with a 3-2-1 point scheme. The scope score characterizes the diversity of the locations of references with respect to the Dewey classification scheme.

The most interesting part of their study is how they calculated atypical connections, which is in essence based on a cocitation measure defined on the Dewey system. A probability matrix was in effect generated. The columns and rows are Dewey decimal codes. The value at $<i, j>$ is the number of times the Dewey decimal codes i and j appeared in the cited references divided by the maximum number of occurrences of the code i. Atypical pairs are defined as those that fall below the lowest 5% of the values. The authors also included other independent variables that are expected to have effects on differentiating high-impact papers, such as prior experience of authors and where and when it was published.

Among six of their logistic regression models, the accuracy of classification ranges from 79.2% to 90.8%. Two models included atypical connections as independent variables. Both models correctly classified 82.5% of papers to either high impact or not high impact. A statistically significant effect of atypical connections was found in one of the two models. The scope seems to be a significant source of impact because the four models that contained the scope variable are 5% more accurate than the two models without it. The same two models included atypical connections instead of scope. The effect of the atypical connections was found to increase the odds of a paper to be high impact by a factor of 15.17.

Schilling and Green's Dewey-based study and Nemet and Johnson's patent classification-based study address the same issue of what is the driving force behind a high-impact idea. Their methodologies, on the other hand, are not comparable because there are differences in almost every aspect of their analyses. What do high-impact papers and highly cited patents have in common? How comparable are Dewey's hierarchy and the U.S. classification system of patents? Nevertheless, it seems to be reasonable and worthwhile to apply Schilling and Green's method to the nine other main classes in the Dewey system. Perhaps we can apply the method to the 500 main class of science. In fact, it is tempting to think about how often atypical links would appear in physical sciences and trigger paradigm shifts. After all, when Thomas Kuhn developed his structure of scientific revolutions, all of his showcase examples were chosen from nature sciences.

6.2.4   Radical Inventions

In a more recent study (published in 2012), Gregory F. Nemet and Evan Johnson investigated whether patents citing prior art patents in different technological domains would attract more patents to cite them than patents that are limited to their own domains (Nemet & Johnson, 2012). This is one instance of the boundary-spanning mechanism characterized in our structural variation theory. It is generally believed that drawing ideas from a diverse range of sources would lead to ideas of higher quality, which would be more likely to be cited. They studied whether the span of a patent's citations could predict the citations to the patent itself. Their study, however, did not find evidence to support this belief. In fact, their evidence seemed contradictory—citing patents in remotely related domains may not help with getting more attention.

In Nemet and Johnson's study, a patent citation refers to a patent A that cites a patent B. The proximity of a patent citation is measured in four levels, near, internal, external, and far external, based on technological domains to which these patents belong. The primary technological domain of a patent is determined by three levels of the U.S. patent classification, namely, the superclass, class, and subclass the patent is assigned to. Nemet and Johnson also considered other potential predictors of patent citations, such as the number of claims in a patent. NB models were used to identify the effects of the proximity on citations in a 10-year window. Since research has shown that citations do not generally follow a normal distribution, NB models are commonly seen in studies of citations.

Nemet and Johnson offered a few explanations for their unexpected results, including the accuracy of the U.S. classification system as a way to define technological domains and their assumption that citation paths indicate paths for the flow of knowledge. Since their study was conducted at the citation level, details such as the context of specific citations and how a prior art patent was actually cited were not investigated in their study. It remains unclear what motivated far external citations and whether near and internal citations were motivated differently. Given the fact that the patenting process is used to establish the novelty and utility of an invention, to what extent could patent citations serve as evidence of incorporating ideas from prior art? To what extent would inventors and patent examiners use patent citations as a mechanism to differentiate a patent from its prior art?

The value of Nemet and Johnson's study is that they clearly articulated the expected role of a boundary-spanning mechanism in driving high-quality patents in terms of their future citations. The weaknesses of the study, as the authors acknowledged, are likely to be caused by the accuracy of the U.S. patent classification system and the accuracy of patent citations.

6.2.5   Genetically Evolved Patentable Ideas

A genetic algorithm (GA) is a search heuristic inspired by the idea of evolution. The heuristic is also called metaheuristics. It is particularly suitable for a wide range of optimization problems. A GA utilizes operations inspired by natural evolution, notably inheritance, mutation, selection, and sexual reproduction (crossover), to generate possible solutions in a given optimization problem. In terms of the fitness landscape metaphor we saw earlier, a GA generates a path for a population to reach a peak in an adaptive landscape. The length of the path is the number of generations between the initial population and the population with the desired fitness level.

A GA typically generates a population of potentially plausible solutions to a problem. These candidate solutions are called individuals (of the population) or phenotypes. The quality of such solutions is measured in terms of their fitness. The population evolves from one generation to the next generation by variations at a lower level of components of an individual, mimicking the mutation, selection, and crossover at the genotype level. At the beginning of an evolution process, the population consists of randomly generated individuals. Individuals with stronger fitness scores will survive to the next generation. The GA iterates a series of generations until a satisfactory solution is reached, enough generations have been tried, proposed solutions are good enough for a human expert to take over the search process, or for other pragmatic reasons.

An interesting application of genetic programming is a series of efforts in duplicating the functionality of designs that have already been patented. The two major criteria of granting patent protection are novelty and utility. Claims in granted patents provide compelling evidence of optimal design decisions made by inventors.

John R. Koza, Martin A. Keane, and Matthew J. Streeter published an intriguing article in *Scientific American* in February 2003 on what genetic programming algorithms can do. They claimed 15 previously patented inventions were duplicated by their GAs. Several of these patents were considered to be seminal in their own fields when they initially appeared. Unfortunately, they didn't provide enough technical details in the *Scientific American* piece, so one has to dig deeper to see what exactly they had achieved.

In 2005, John R. Koza, Sameer H. Al-Sakran, and Lee W. Jones provided more detail in a paper presented at Genetic and Evolutionary Computation Conference (GECCO). Their goal was to demonstrate that their genetic programming approach could achieve a level of intellectual competence comparable to human intelligence. They argued that if their genetic programming approach could replicate something that is independently evaluated and endorsed, then they would have a compelling case. Patents granted in the past provide a good source of such cases. The types of problems particularly suitable for this approach are optimization problems, for example, the design of an optical system such as a telescope eyepiece.

The design of a telescope eyepiece is an optimization process. Recall that in Chapter 5 we related the evolution process to the design of a telescope. The design of a telescope eyepiece typically needs to specify the configuration of a number of lenses of various curvatures and other properties (such as the distance between lenses). The overall quality of a configuration is measurable, thus improvements can be made by making various adjustments in the configuration of each element in the system. The quality of a configuration is considered its fitness value. Different configurations represent individuals from the same or different populations, depending on how they differ along each of the multiple dimensions. Choosing the best configuration is, therefore, an optimization process or an adaptive walk over an adaptive landscape to reach the highest adaptive peak. The significance of the arguments made by John Koza and his colleagues is that the adaptive walk is capable of reaching altitudes as high as where human inventors have reached in the past. An even stronger statement is that the generic adaptive walk can reach the fitness level as high as where human inventors have reached in the recent past. Along this line of reasoning, one may wonder whether it will become feasible for the genetic programming approach to reach a level of fitness that no human inventors have ever been there.

In their GECCO'2005 paper, Koza et al. described how genetic programming was used as an "invention machine" and automatically synthesize complete designs for six optical lens systems. The design of an optical lens system is complex. The designers need to make numerous decisions to produce the desirable final outcome, that is, the fittest phenotype. Making decisions is making selections from a space of candidate elements. For example, how many lenses will be used in the system? How will these lenses be arranged? What are the best choices of properties for each individual lens? What about the shape, curvature, and size of each lens? Echo individual design in the population is a combination of such multiple components. The resemblance to the problems faced by evolutionary genetics is striking. The potential search space is huge. There are numerous potentially valid combinations. On the other hand, the number of really valuable combinations is much smaller than the entire search space. Searching needles in a haystack would still be an accurate way to characterize the nature of the complexity. The topologies of such search spaces are usually unknown. No one knows how many peaks are in the high-dimensional space. No one knows how far away they are located from each other. How many generations of evolution would it take to reach another seemingly more promising peak? The experiment by Koza and his colleagues provided some concrete information for us to glimpse into the types of answers we might get in general.

Bear in mind, designing a telescope is by no means simple, but there are many inventions that have the complexity of several magnitudes higher, for example, the design of an airplane.

Conventionally, an optical lens system is specified by a prescription, which is a table that summarizes components of the system and their properties. For example, a prescription specifies the number of lenses, the radius of each lens, the materials at particular positions, and the aperture. Koza et al. emphasized that their genetic programming search started from scratch. In terms of an adaptive landscape, their search could start from the bottom of a valley to climb up all the way to approach some peaks. Once a prescription is given for an optical system, many of its optical properties can be calculated by ray-tracing analytic and simulation techniques. The path of how light rays of various wavelengths run through the system can be computationally simulated so that designers are informed about the characteristics of the system in terms of distortion, astigmatism, and chromatic aberration.

In the space of all the potential valid designs, the design process of a telescope eyepiece can be automated to the following procedure:

1. Move to a point on a straight line.
2. Pick a lens of some sort.
3. Place the chosen lens at the current point.
4. Move to another point further along the line.
5. Repeat steps 2–4 to place as many lenses as desirable.
6. Measure the fitness of the individual design.

Each step may involve selections at a finer level of granularity, such as the curvature and the size of the lens, and the distance between adjacent lenses.

The fitness measure of a particular configuration of the optical system typically reflects a combinatory effect of its numerous components and their properties. The performance of the system can be characterized by ray-tracing outcomes with respect to desirable performance measures and physical constraints. Physically infeasible configurations would be penalized in measuring the fitness. For example, the image should always fall outside all the lenses. Koza et al. estimated that the size of the design population was about 75,000. They applied the genetic programming generator to six patented eyepieces between 1940 and 2000.

The evolution became similar to a 1953 patent in the 257th generation of the population. The genetic programming solution recommended the same number of lenses, surfaces, and one doublet and three singlet lenses. For a 2000 patent, the search in the 295th generation was considered to be close enough in terms of functionality because, interestingly, the genetic search found a different way from the patent to achieve the functionality. For a 1985 patent, the best design in the 300th generation had a performance comparable to the patented design.

Koza et al. explained in more detail about of the six patents—the Konig and Tackaberry–Muller eyepiece. The best genetically evolved design in the 490th generation was identified as comparable to the functionality of the Tackaberry–Muller patent, which cited the 1940 Konig patent. The fittest individual in the 490th generation was slightly better than the Tackaberry–Muller patent. Koza et al. claimed that the genetically evolved design would infringe the first claim of the Konig patent. In other words, they claimed that their genetic programming approach evidently can find a design solution as good as what a human inventor can do. They concluded that genetic programming is capable of producing human-competitive solutions to complex problems.

6.3 BRIDGING THE GAPS

Critical transitions in a CAS may have early signs that one can detect. In the context of the growing scientific knowledge, critical transitions correspond to the type of scientific revolutions characterized by Thomas Kuhn. The structural variation theory provides a framework for us to identify the potential value of newly available information. The studies that modeled how radical ideas may emerge in patents underscore the common theme of a recombinant search in a high-dimensional space. In this section, we will describe the implementation of the structural variation theory. Examples of how we applied this procedure are provided in next section.

6.3.1 The Principle of Boundary Spanning

A recurring theme from a diverse body of work is that bridging gaps between previously disjoint bodies of knowledge plays a profound role in creativity. Notable studies include the work of Ronald S. Burt in sociology (Burt, 2004), Donald Swanson in information science (Swanson, 1986a), and conceptual blending as a theoretical framework for exploring human information integration (Fauconnier & Turner, 1998). We have developed an explanatory and computational theory of transformative discovery based on criteria derived from structural and temporal properties (Chen, 2011; Chen et al., 2009).

In the history of science, there are many examples of how new theories revolutionized the contemporary knowledge structure. For example, the 2005 Nobel Prize in Medicine was awarded to the discovery of *Helicobacter pylori*, a bacterium that was previously thought not possible to find in the human gastric system (Chen et al., 2009). In literature-based discovery, Swanson discovered a previously unnoticed linkage between fish oil and Raynaud's syndrome (Swanson, 1986a). In terrorism research, before the September 11 terrorist attacks, it was widely believed that only those who directly witnessed a traumatic scene or directly experienced trauma could be at risk for PTSD. Later research, however, has shown that people may develop PTSD simply by watching the coverage of a traumatic scene on TV (Chen, 2006). In drug discovery, one of the major challenges is to find new compound structures effectively in the vast chemical space that satisfy an array of constraints (Lipinski & Hopkins, 2004). In these and many more scenarios, a common challenge for coping with a constantly changing environment is to estimate the extent to which the structure of a network should be updated in response to newly available information.

Many studies have addressed factors that could explain or even predict future citations of a scientific publication (Aksnes, 2003; Hirsch, 2007; Levitt & Thelwall, 2008; Persson, 2010). For example, is a paper's citation count from last year a good predictor for new citations this year? Are the download times a good predictor of citations? Is it true that the more references a paper cites, the more citations it will receive later on? Similarly, the potential role of prestige, or the Matthew Effect (coined by Robert Merton), has been commonly investigated, ranging from the prestige of authors to the prestige of journals in which articles are published (Dewett & Denisi, 2004). Many of these factors, however, are loosely and indirectly coupled with the conceptual and semantic nature of the underlying subject matter of concern. We will refer to them as extrinsic factors. In contrast, intrinsic factors have direct and profound connections with the intellectual content and structure. One example of an intrinsic factor is concerned with the structural variation of a field of study. A notable example is the work Swanson did on linking previously disjoint bodies of knowledge, such as the connection between fish oil and Raynaud's syndrome (Swanson, 1986a).

The theoretical underpinning of the structural variation is that scientific discoveries, at least a subset of them, can be explained in terms of boundary-spanning, brokerage, and synthesis mechanisms in an intellectual space (Chen et al., 2009). This conceptualization generalizes the principle of literature-based discovery pioneered by Swanson (Swanson, 1986a, 1986b), which assumes that connections between previously disparate bodies of knowledge are potentially valuable. In Swanson's famous ABC model, the relationships AB and BC are known in the literature. The potential relationship AC becomes a candidate that is subject to further scientific investigation (Weeber, 2003). Our conceptualization is more generic in several ways. First, in the ABC model, the AC relation changes an indirect connection to a direct connection, whereas our structural variation model makes no assumption about any prior relations at all. Second, in the ABC model, the scope of consideration is limited to relationships involving three entities. In contrast, our structural variation model takes a wider context into consideration and addresses the novelty of a connection that links groups of entities as well as connections linking individual entities. Because of the broadened scope of consideration, it becomes possible to search for candidate connections more effectively. In other words, given a set of entities, the size of the search space of potential connections can be substantially reduced if additional constraints are applicable for the selection of candidate connections. For example, the structural hole theory developed in social network analysis emphasizes the special potential of nodes that are strategically positioned to form brokerage, or boundary spanning, links and create good ideas (Burt, 2004; Chen et al., 2009).

6.3.2 Baseline Networks

We expect that the degree of structural variation introduced by a new article can offer prospective information because of the boundary-spanning mechanism. If an article introduces novel links that span the boundaries of different topics, then we expect these will signify its potential in taking the intellectual structure for a new turn. We model the intellectual structure of scientific knowledge in terms of a network of cocited references in a chosen domain. More than a decade ago, Henry Small demonstrated the concept-symbol role played by cited references. More specifically, a reference tends to have an emergent property to become a symbolic representation of a more intangible concept based on the way the reference has been cited in various publications. The network representation of the current knowledge therefore provides a baseline for us to assess the novelty of a newly proposed idea.

Each publication cites a set of references. Citations of these references are compared with the structure of the baseline network so that unprecedented links and reinforcement links can be identified. The transformative potential of the article will be measured by metrics derived from these links. We will refer to these metrics as structural variation metrics. The predictive effects of these structural variation metrics will be estimated by NB models and zero-inflated negative binomial (ZINB) models. The dependent variable is the number of times an article was cited. The independent variables include the structural variation metrics and a few other variables, such as the number of coauthors and the number of references cited by the article in question. More specific details will be explained in context.

The basic assumption in the structural variation theory is that the extent of a departure from the current intellectual structure is a necessary condition for a potentially transformative idea in science. A potentially transformative idea needs to bring changes to the existing structure of knowledge in the first place. In order to measure the degree of structural variation introduced by a scientific article, the intellectual structure at a particular moment of time needs to be represented in a way such that structural changes can be computationally detected and manually verified.

A network represents how a set of entities are connected. Entities are represented as nodes, or vertices, in the network. Their connections are represented as links or edges. Relevant entities in our context include several types of information that can be computationally extracted from a scientific article, such as references cited by the article, authors and their affiliations, the journal in which the article is published, and keywords in the article. Here we will limit our discussions to networks that are formed with a single type of entities, although networks of multiple types of entities are worth considering once we establish a basic understanding of structural variations in networks of a single type of entities.

Networks of co-occurring entities represent a wide variety of types of connectivity. A network of co-occurring words represents how words are related in terms of if and how often they appear in the vicinity of each other. Cocitation networks of entities such as references, authors, and journals can be seen as a special case of co-occurring networks. For example, cocitation networks of references are networks of references that appear together in the bodies of scientific papers—these references are cocited.

Networks of cocited references represent more specific information than networks of cocited authors because references of different articles by the same author would be lumped together in a network of cocited authors. Similarly, networks of cocited references are more specific than networks of cocited journals. We refer such differences in specificity as the granularity of networks. Measurements of structural variation need to take the granularity factor into account because it is reasonable to expect that networks at different levels of granularity would lead to different measures of structural variations.

Another decision to be made about a baseline network is a sampling issue. Taking a particular year as a standing point to look at in the past, how far back should we consider the construction of a baseline network that would adequately represent the underlying intellectual structure? Does the network become more accurate if we go back further into the past? Will it be more efficient if we limit it to the most recent years that really matter the most? Given articles published in a particular year Y_i, the baseline network represents the intellectual structure using information from articles published up to year Y_{i-1}. Two types of baseline networks are feasible in this context: one is to use a moving window of a fixed size $[Y_{i-k}, Y_{i-1}]$ and the other is to use the entire history $(Y_0, Y_{i-1}]$, where Y_0 is the earliest year of publication for records in the given dataset.

6.3.3   Structural Variation Metrics

Given the boundary-spanning principle, one may derive numerous metrics to measure the changes caused by contrasting with another set of links. We will illustrate the development of three structural variation metrics, namely, modularity change rate (MCR), cluster linkage (CL), and centrality divergence (C_{KL}). The first two metrics are defined on the basis of a decomposition of the baseline network, that is, how links are distributed over a set of clusters of nodes. The third one does not require a decomposition of the baseline network.

Modularity Change Rate

Suppose the baseline network is decomposed into a number of clusters of nodes. A decomposition is also known as a partition. The quality of a partition can be measured by the modularity of the network with respect to the partition. The modularity measures the extent to which connections between different clusters are minimized by the partition. If different clusters are loosely connected, then the overall modularity would be high. In contrast, if clusters are interwoven, then the modularity would be low. Newman's algorithm (Newman, 2006) is a good choice to calculate the modularity with reference to a cluster configuration generated by spectral clustering (Chen, Ibekwe-SanJuan, & Hou, 2010; Luxburg, 2006). Figure 6.5 shows an example of the modularity measures of a series of cocitation networks of research in regenerative medicine. The modularity was relatively stable over the years, except in 2007 and 2010. As it turned out, the substantial drops of modularity were caused by research on induced pluripotent stem cells (iPSCs), which was awarded the Nobel Prize in 2012.

Suppose the network G is partitioned by a partition C into k clusters such that $G = c_1 + c_2 + \cdots + c_k$; the modularity $Q(G)$ is defined as follows, where m is the total number of edges in the network G and n is the number of nodes in G. $\delta(c_i, c_j)$ is known as the Kronecker delta. It is 1 if nodes n_i and n_j belong to the same cluster and 0 otherwise. $\deg(n_i)$ is the degree of node n_i. The range of $Q(G)$ is between -1 and 1:

$$Q(G,C) = \frac{1}{2m} \sum_{i,j=0}^{n} \delta\left(c_i, c_j\right) \cdot \left(A_{ij} - \frac{\deg\left(n_i\right) \cdot \deg\left(n_j\right)}{2m} \right)$$

The modularity of a network is a measure of the overall structure of the network. Its range is between -1 and 1. The MCR of a scientific paper measures the relative structural change due to the information from the published paper with reference to

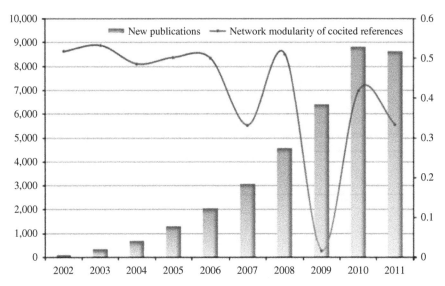

FIGURE 6.5   Modularity measures changed in a series of cocitation networks of regenerative medicine.

a baseline network. For each article a and a baseline network G_{baseline}, the MCR is defined as follows:

$$\text{MCR}(a) = \frac{Q\left(G_{\text{baseline}},C\right) - Q\left(G_{\text{baseline}} \oplus G_a,C\right)}{Q\left(G_{\text{baseline}},C\right)} \cdot 100$$

where $G_{\text{baseline}} \oplus G_a$ is the updated baseline network by information from the article a. For example, suppose reference nodes n_i and n_j are not connected in a baseline network of cocited references but they are cocited by article a; a new link between n_i and n_j will be added to the baseline network. In this way, the article changes the structure of the baseline network.

Intuitively, adding a new link anywhere in a network should not increase the modularity of the network. It should either reduce it or leave it intact. The change in modularity, however, is not a monotonic function like we initially thought. In fact, it depends on where the new link is added and how the network is structured. Adding a link may reduce the proportion of the modularity in some clusters, but may increase the modularity of other clusters in the network. Thus, the overall modularity change is not monotonic.

In summary, the updated modularity may increase as well as decrease, depending on the structure of the network and where the new link is added. With this particular definition of modularity, between-cluster links are always associated with a zero-valued term in the overall modularity formula due to the Kronecker delta. What we see in the change in modularity is a combination of results from several scenarios that are indirectly affected by the newly added link. We will introduce our next metric to reflect the changes in terms of between-cluster links directly.

Cluster Linkage

The CL measures the overall structural change introduced by an article a in terms of new connections added between clusters. Its definition assumes a partition of the network. We introduce a function of edges $\lambda(c_i, c_j)$, which is the opposite of δ_{ij} used in the modularity definition. The value of λ_{ij} is 1 for an edge across distinct clusters c_i and c_j. It will be 0 for edges within a cluster. λ_{ij} will allow us to concentrate on between-cluster links and ignore within-cluster links, which is the opposite of how the modularity metric is defined. The new metric *linkage* is the sum of all the weights of between-cluster links e_{ij} divided by K—the total number of clusters in the network. A cluster cannot link to itself, that is, we assume $e_{ii}=0$ for all nodes. Using link weights makes the metric sensitive to links that strengthen connections between clusters in addition to novel links that make unprecedented connections between clusters.

It is possible to take into account the size of clusters that a link is connecting so that connections between larger-sized clusters become more prominent in the measurement. For example, one option is to multiple each e_{ij} by $\sqrt{\text{size}(c_i) \cdot \text{size}(c_j)} / \max(\text{size}(c_k))$. Here the metric is defined without such modifications for simplicity. If C is a partition of G, then the *linkage* metric is defined as follows:

$$\text{Linkage}(G,C) = \frac{\sum_{i \neq j}^{n} \lambda_{ij} e_{ij}}{K}$$

$$\lambda_{ij} = \begin{cases} 0, n_i \in c_j \\ 1, n_i \notin c_j \end{cases}$$

The CL is defined as the difference of *linkage* before and after new between-cluster links added by an article a:

$$\text{CL}(a) = \Delta\text{Linkage}(a) = \text{Linkage}\left(G_{\text{baseline}} \oplus G_a, C\right) - \text{Linkage}\left(G_{\text{baseline}}, C\right)$$

Linkage($G+\Delta G$) is always greater than or equal to Linkage(G). Thus, CL is nonnegative.

Centrality Divergence

The C_{KL} metric measures the structural variation caused by an article a in terms of the divergence of the distribution of betweenness centrality $C_B(v_i)$ of nodes v_i in the baseline network. This definition does not involve any partitions of the network. If n is the total number of nodes, the degree of structural change $C_{\text{KL}}(G, a)$ can be defined in terms of the K–L divergence:

$$C_{\text{KL}}\left(G_{\text{baseline}}, a\right) = \sum_{i=0}^{n} p_i \cdot \log\left(\frac{p_i}{q_i}\right)$$

$$p_i = C_B\left(v_i, G_{\text{baseline}}\right)$$

$$q_i = C_B\left(v_i, G_{\text{updated}}\right)$$

For nodes where $p_i = 0$ or $q_i = 0$, we reset them as a small number, such as 10^{-6}, to avoid log(0).

6.3.4   Statistical Models

We will use NB and ZINB models to validate the role of structural variation in predicting future citation counts of scientific publications. The NB distribution is generated by a sequence of independent Bernoulli trials. Each trial is either a "success" with a probability of p or a "failure" with a probability of $(1-p)$. Here, the terminology of success and failure in this context does not necessarily represent any practical preferences. The random number of successes X before encountering a predefined number of failures r has an NB distribution:

$$X \sim NB(r, p)$$

One can adapt this definition to describe a wide variety of count events. Citation counts belong to a type of count events with an overdispersion, that is, the variance is greater than the mean. NB models are commonly used in the literature to study this type of count events. Two types of dispersion parameters are used in the literature, θ and α, where $\theta \cdot \alpha = 1$.

Zero-inflated count models are commonly used to account for excessive zero counts (Hilbe, 2011; Lambert, 1992). Zero-inflated models include two sources of zero citations: the point mass at zero $I_{\{0\}}(y)$ and the count component with a count distribution f_{count}(counts) such as NB or Poisson (Zeileis, Kleiber, & Jackman, 2011). The probability of observing a zero count is inflated with probability $\pi = f_{zero}$(zero citations):

$$f_{zero-inflated}(\text{citations}) = \pi \times I_{\{0\}}(\text{citations}) + (1 - \pi) \times f_{count}(\text{citations})$$

ZINB models are increasingly used in the literature to model excessive occurrences of zero citations (Fleming & Bromiley, 2000; Upham, Rosenkopf, & Ungar, 2010). The report of a ZINB model consists of two parts: the count model and the zero-inflated model. The Vuong test is one way to test the superiority of a ZINB model compared to a corresponding NB model. The Vuong test is designed to test the null hypothesis that the two models are indistinguishable. Akaike's Information Criterion (AIC) is also commonly used to evaluate the viability of a model. Models with lower AIC scores are regarded as better models.

Global citation counts of scientific publications recorded in the Web of Science are used in the following NB models with the logarithm link function.

Global citations ~ Coauthors + Modularity Change Rate + Cluster Linkage + Centrality Divergence + References + Pages

Global citations is the dependent variable. *Coauthors* is a factor of three levels of 1, 2, and 3. Level 3 is assigned to articles with three or more coauthors. *Coauthors* is an indirect indicator of the extent to which an article synthesizes ideas from different areas of expertise represented by each coauthor.

Three structural variation metrics are included as covariants in generalized linear models, namely, MCR, CL, and C_{KL}. According to our theory of creativity, groundbreaking ideas are expected to cause strong structural variations. If global citation counts provide a reasonable proxy of recognitions of intellectual contributions in a scientific community, we would expect that at least some of the structural variation metrics will have statistically significant main effects on global citations.

The number of cited references and the number of pages are commonly reported in the literature as good predictors of citations. In order to compare the effects of structural variation with these commonly reported extrinsic properties of scientific publications, *References* and *Pages* are included in the models. Our theory offers a simpler explanation for why the more references a paper cites, the more citations it appears to get. Due to the boundary-spanning synthetic mechanism, an article needs to explain multiple parts and how they can be innovatively connected. This process will result in citing more references than an article that covers a narrower range of topics. Review papers by nature belong to this category.

It is known that articles published earlier tend to have more citations than articles published later. The exposure time of an article is included in the NB models in terms of a logarithmically transformed year of publication of an article.

An intuitive way to interpret coefficients in NB models is to use incidence rate ratios (IRRs) estimated by the models. For example, if *coauthors* has an IRR of 1.5, it means that as the number of coauthors increases by 1, the global citation counts would be expected to increase by a factor of 1.5, that is, increasing 1.5 times, while holding other variables in the model constant. In our models, we will particularly examine statistically significant IRRs of structural variation models.

ZINB models use the same set of variables. The count model of ZINB is identical to the NB model described earlier. The zero-inflated model of ZINB uses the same set of variables to predict the excessive zeros. We found little in the literature about good predictors of zeros in a comparable context. We choose to include all the six variables in the zero-inflated model to provide a broader view of the zero-generating process. ZINBs are defined as follows:

Global citations ~ Coauthors + Modularity Change Rate + Cluster Linkage + Centrality Divergence + References + Pages

Zero citations ~ Coauthors + Modularity Change Rate + Cluster Linkage + Centrality Divergence + References + Pages

6.4 APPLICATIONS

We will demonstrate the applications of the structural variation theory with three examples. The first example is an analysis of the evolution of research in small-world networks. The second example is the evolution of complex network analysis, which had become an emerging field in the early 2000s. In the third example, we will apply the same procedure to a study of a collection of patents.

6.4.1 Small-World Networks

Duncan Watts and Steve Strogatz (1998) published their groundbreaking article that subsequently attracted researchers from a wide range of disciplines, from social sciences to physical sciences. Since Watt and Strogatz's article had evidently played a transformative role in the rapid development of complex network research, we will focus on how the structure of relevant knowledge changed in the period of time in which the pivotal article was published.

We started with Watt and Strogatz's 1998 article as a seed to construct a dataset of articles. The seed article was already highly cited. In 2011, it had over 4660 citations on the Web of Science. According to Google Scholar, the article was cited by 21,313 documents on the Internet in 2013.

Our dataset consisted of articles that can be considered peers of the seed article and articles that cited the seed article. The peer articles include articles that shared one or more cited references with the seed article. In addition, these articles may cover a broader range of topics than those addressed by the seed article and thereby provide a broader context where we can examine the novelty of structural changes induced by new articles as well as by the seed article. The resultant dataset included 5135 articles published between 1990 and 2010.

Figure 6.6 shows a network of cocited references derived from the 5135-article dataset. The network contained 205 references and 1164 cocitation links. The network was partitioned to 12 clusters, which are shown in the visualization as areas of concentrations of cited references. Each cluster was automatically labeled by terms extracted from the context in which members of the cluster were cited. The modularity of the partition was 0.6537, and the mean silhouette score was 0.811. The level of modularity indicates that the components in the network are reasonably separated with a few connections between clusters. The mean silhouette score of 0.811 suggests that these clusters are highly homogeneous.

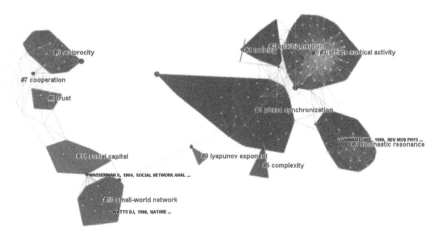

FIGURE 6.6   The synthesized network of research in small-world network.

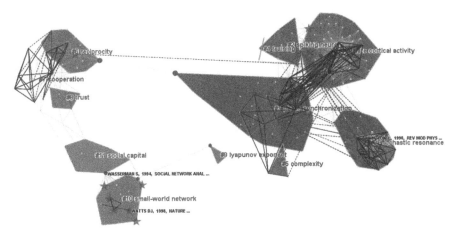

FIGURE 6.7 A network of cocited references derived from 5135 articles published on small-world networks between 1990 and 2010. The network of 205 references and 1164 cocitation links is divided into 12 clusters with a modularity of 0.6537 and the mean silhouette of 0.811. The red lines are made by the top-15 articles with the largest centrality variation rate. (*See insert for color representation of the figure.*)

Cluster 10, labeled as "small-world network," in the lower left corner of the visualization was the island emerged after the publication of the seed article. The island contains the seed article and the references of other articles. Its nearest neighbor is cluster 11, labeled as "social capital." A reference to Wasserman's 1994 book *Social Network Analysis* in cluster 11 was marked by a purple circle, indicating its position as a "gatekeeper" between the two clusters. Cluster 11 represents previous research in small-world networks in social sciences.

In Figure 6.7, we superimposed the visualization with cocitation links from a group of articles that ranked the highest in terms of their structural variation potential. We calculated the centrality variation rate of each article in the dataset and chose the top 15 articles according to this score. According to the structural variation theory, these articles contributed most to the system-level changes. In particular, these articles had caused most of changes in the distribution of betweenness centrality scores of the references.

The distribution of the superimposed links highlighted that they were mostly bridges between different islands of cited references. Most of the 15 articles were evidently significant in their own right because they also appeared on the visualization marked by the symbols of stars. Interestingly, 6 of the 15 articles were on the same island as the seed article. Two stars appeared on the island of cluster 3, which was labeled as "striate cortical activity." One of them appeared on the island of cluster 6, labeled as "reciprocity."

6.4.2 Complex Network Analysis (1996–2004)

This dataset overlaps with the small-world networks dataset, but the two datasets were constructed differently. A topic search for complex network analysis revealed two of the most frequently cited articles: those of Barabasi and Albert (1999) and

Watts and Strogatz (1998). We constructed an updated dataset with these two articles as seed articles. First, one seed article was used to form a subdataset. Then, the two datasets were merged to form a combined dataset on complex network analysis.

The seed article (Barabasi & Albert, 1999) has the highest citation count. As of August 5, 2011, it has been cited 5792 times in the Web of Science. The article of Watts and Strogatz (1998) has the second highest citation count of 5291. For each of the seed articles, we retrieved all the records that share at least one cited reference with the seed article. Then, we merged all the records and formed a combined dataset. The seed Barabasi_1999 has 8919 related articles published between 1980 and 2011. If we limited the period to 1996–2004 and original research articles, reviews, and proceedings, 2326 records would be retained in the subset. The seed article Watts_1998 has 14,393 related records. Interestingly, this is 1.6 times more than what Barabasi_1999 has, although Barabasi_1999 has 500 more citations. It appears that Watts_1998 involves a wider range of topics than the Barabasi_1999 article. The merged dataset from the two subsets contains 6764 records. We first analyzed the dataset seeded by Barabasi_1999 and then the dataset seeded by Watts_1998, and finally, the merged dataset.

ZINB regression models were used to estimate the effects of variables more accurately, especially in situations where many cases may have a value of zero. It is known that a considerable number of scientific publications would never be cited. Thus, using ZINBs would be appropriate in our case. In fact, we generated an NB model as well as a ZINB model. The Vuong test indicated that the ZINB model is superior to the NB model at the p-value of 0.0033. In the following discussions, we will focus on the ZINB model.

As shown in Table 6.3, the dependent variable in the ZINB model was the global citation count of 3515 citing articles on complex network analysis, published between 1996 and 2004. The global citation count is the value of the Times Cited field of a record in the Web of Science at the time of data retrieval. Independent variables in the model included structural variation metrics, namely, MCR, weighted CL, and C_{KL}, and nonstructural metrics such as the numbers of coauthors, references, and pages. Effects estimated by the ZINB model are listed in the few columns in the middle of the table, where effects estimated by the NB model are listed in the two rightmost columns.

Under the ZINB model, the first column below the heading of count model lists the estimates of IRRs of the independent variables. The next column lists the corresponding p-levels. Statistically significant results are highlighted in a bold type. The IRR of a variable indicates as the value of the variable increases by 1, how many times by which the value of the dependent variable would be expected to increase. The IRR of coauthors is 1.293, which means that for an article with one more coauthor, its global citation count would be expected to increase by 1.293 times. The predictive effect of weighted CL on global citations was the strongest, with an IRR of 3.103. As the weighted CL increased by one unit, the global citations would increase by 3.103 times. The MCR was found to have a minor predictive effect on global citations, with an IRR of 1.080. In comparison, nonstructural metrics did not have any substantial impact on global citations. The NB model found a similar pattern. The weighted CL had the strongest effect, with an IRR of 3.160.

Table 6.3 ZINB regression and NB models of global citation counts of 3515 citing articles on complex network analysis (1996–2004)

Global cites	ZINB					
	Count model Negbin with log link		Zero-inflation model Binomial with logit link		NB	
Coauthors	**1.293**	**0.000**	0.062	0.077	**1.306**	**0.000**
Modularity change rate	**1.080**	**0.014**	**0.012**	**0.044**	**1.083**	**0.025**
Weighted cluster linkage	**3.103**	**0.000**	1.304	0.906	**3.160**	**0.000**
Centrality Divergence	0.391	0.237	385363.6	0.102	0.343	0.184
No. of references	**1.013**	**0.000**	**0.489**	**0.027**	**1.013**	**0.000**
No. of pages	**0.970**	**0.000**	1.133	0.120	**0.970**	**0.000**
Dispersion parameter (θ)	0.536				0.528	
AIC			31.768			31.787
Vuong test (ZINB > NB)			-2.7186, $p = \mathbf{0.0033}$			

AIC, Akaike's Information Criterion.
Coefficients are incidence rate ratios. Weighted cluster linkage is the strongest predictor of citation counts, followed by the number of coauthors, and the modularity change rate. In this case, with a lower AIC and a statistically significant Vuong test, the ZINB model is superior.

Based on the results, we can conclude that the structural variation mechanisms are evident in statistical models of global citation counts. Structural variation metrics such as CL are better predictors of global citation counts than more commonly studied factors such as the number of references. As a result, the structural variation paradigm offers simpler interpretations of why a number of factors appear to be associated with high citations. Broadly speaking, CL measures the boundary-spanning strength of an idea; C_{KL} measures the transformative potential of an idea; and MCR measures changes in network modularity.

6.4.3 National Cancer Institute's Patent Portfolio

The availability of the abundant patent data, the increasingly widespread awareness of information visualization, and the maturity of search engines on the Web are among the most influential factors behind the emerging trend of patent analysis. Many patent search interfaces allow users to search by specific sections in patent databases, for example, by claims. Statistical analysis and intuitive visualization functions are by far the most commonly seen selling points from a salesman's patent analysis portfolio. The term visualization becomes so fashionable now in the patent analysis industry that from time to time we come across visualization software tools that turn out to be little more than standard displays of statistics.

Numerous newly formed companies are specifically aiming at the patent analysis market. Apart from historical driving forces such as monitoring knowledge and technology transfer and staying in competition, the rising commercial interest in patent analysis is partly due to the public accessible patent databases, notably the huge amount of patent applications and grants from the USPTO. The public can search

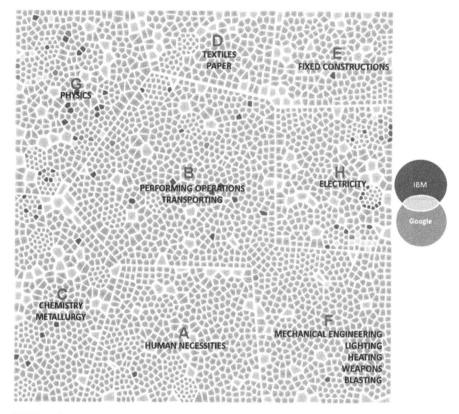

FIGURE 6.8   A Voronoi diagram of patents by IPC. Red, IBM; green, Google; yellow, both. (*See insert for color representation of the figure.*)

patents and trademarks at USPTO's website http://www.uspto.gov/ and download bibliographic data from ftp://ftp.uspto.gov/pub/patdata/. These public records provide a valuable source for the study of patenting activities. Figure 6.8 shows an overview of a density landscape of patents. The entire base map is divided into eight smaller areas A–H according to the International Patent Classification (IPC). These areas are in turn divided further into subclasses of IPC. One can superimpose different groups of patents over the base map and visualize various patterns along the IPC classification. In Figure 6.8, subclasses containing IBM patents are shown in red, whereas those containing Google patents are shown in green. Subclasses in which the two groups overlap are shown in yellow. IBM patents appeared in seven of the eight areas. Google patents appeared in four of them. They overlap in three areas, primarily in physics (G) and electricity (H).

The National Cancer Institute (NCI) has been conducting a variety of cancer research projects. Patents invented by NCI researchers represent an important intellectual asset. One would ask a similar set of questions about NCI's patent portfolio as we did with the literature of research fields, such as complex network analysis,

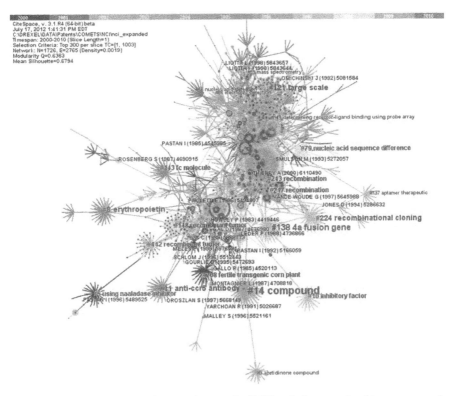

FIGURE 6.9 A minimum spanning tree of a network of 1726 cocited patents related to cancer research. (*See insert for color representation of the figure.*)

mass extinctions, and terrorism research. Are there radical patents in the portfolio? What are the predictive effects of structural variation metrics on patent citations?

We constructed a dataset of patents based on an expanded version of the COMETS patent database. First, all the patents with NCI inventors were identified and used to form the initial core of the dataset. Second, we retrieved all the U.S. patents that cited at least one patent in common with what the NCI patents cited. These patents would provide the context that covers the NCI patents as a whole. Figure 6.9 shows a visualization of a network of 1726 patents cited by the final dataset. The visualized network represents the citation footprint of patents granted between 2000 and 2010. The citing behavior of top 300 most cited patents each year was used to generate the synthesized network. The synthesized network was then partitioned into a number of patent cocitation clusters, including cluster 14 labeled as "compound" at the lower part of the visualization and clusters 213 and 217 in the center of the visualization with a label of "recombination." Patents with red rings have patent citation bursts—suddenly increased citations.

In the following example, we applied the structural variation theory to the analysis of transformative potentials of patents. The same procedure was used, namely, the ZINB and NB models with structural variation metrics to predict patent citations.

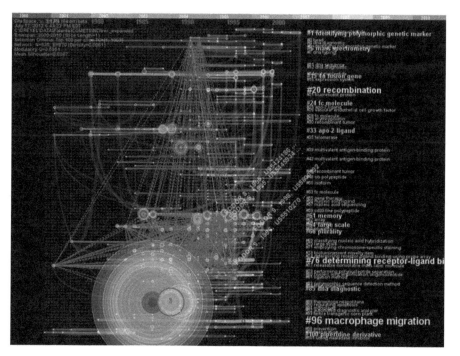

FIGURE 6.10   A timeline visualization of a broader context of the NCI patents. Dashed lines represent transformative links connecting different clusters of patents. Patents in each cluster are shown horizontally by their granted date from left to right. Labels next to clusters on the right characterize the primary topics of impact. (*See insert for color representation of the figure.*)

Figure 6.10 shows a timeline visualization of the patents cited in a broader context of the patents expanded from the NCI patents. The timeline visualization depicts over 100 clusters of patents grouped by patent cocitation links. Clusters are displayed horizontally from the earliest patents on the left to the most recent patents on the right. The label of a cluster was chosen from patents that cited members of the cluster to characterize the nature of the impact. For example, cluster 96 was labeled with the term *macrophage migration*, which means the majority of the citations to the cluster came from patents on "macrophage migration." The density of the dashed lines suggests that most of the transformative links were added between clusters from 61 through 79. Long-range transformative links were added between cluster 3 and the group of clusters in the middle of the display.

Figure 6.11 illustrates the calculation of patent cocitation strengths in detail. The most cited NCI patent in our dataset was U.S. Patent 4676980, entitled "target specific cross-linked heteroantibodies." We will denote the patent as "P." Patent P's application was filed in 1985 and granted in 1987. The patent has six claims. It was cited by 197 patents in the expanded dataset. The patent cited three patents, A, B, and C, which were granted earlier, one in 1977 and two in 1984. Multiple IPC classes were assigned to each of the four patents, A–C and P.

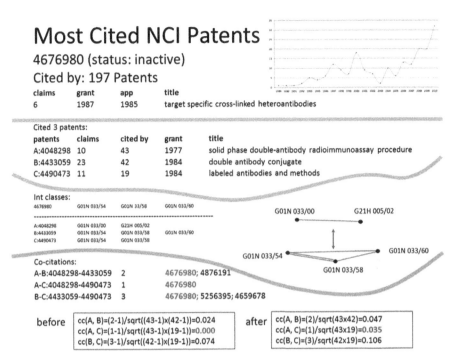

Most Cited NCI Patents

4676980 (status: inactive)
Cited by: 197 Patents

claims	grant	app	title
6	1987	1985	target specific cross-linked heteroantibodies

Cited 3 patents:

patents	claims	cited by	grant	title
A:4048298	10	43	1977	solid phase double-antibody radioimmunoassay procedure
B:4433059	23	42	1984	double antibody conjugate
C:4490473	11	19	1984	labeled antibodies and methods

Int classes:

4676980	G01N 033/54	G01N 33/58	G01N 033/60
A:4048298	G01N 033/00	G21H 005/02	
B:4433059	G01N 033/54	G01N 033/58	G01N 033/60
C:4490473	G01N 033/54	G01N 033/58	

Co-citations:

A-B:4048298-4433059	2	4676980; 4876191
A-C:4048298-4490473	1	4676980
B-C:4433059-4490473	3	4676980; 5256395; 4659678

before	after
cc(A, B)=(2-1)/sqrt((43-1)x(42-1))=0.024	cc(A, B)=(2)/sqrt(43x42)=0.047
cc(A, C)=(1-1)/sqrt((43-1)x(19-1))=0.000	cc(A, C)=(1)/sqrt(43x19)=0.035
cc(B, C)=(3-1)/sqrt((42-1)x(19-1))=0.074	cc(B, C)=(3)/sqrt(42x19)=0.106

FIGURE 6.11 The impact of patent 4676980 on the existing network structure. Before patent 4676980 was granted, patents A and C were not cocited at the patent level. After it was granted, patents A and C became cocited with a strength of 0.035.

Three IPC classes, G01N 033/54, 33/58, and 33/60, were assigned to patent P. Before patent P was granted, patents A and C were not cocited at the patent level. After it was granted, the cocitation strength between patents A and C became 0.035. At the IPC level, patent P's citations formed two groups of IPC classes. One group contained G01N 33/00 and G21H 005/02. The other group contained G01N 033/54, 033/58, and 033/60.

Figure 6.12 shows whether or not patent P introduced structural changes at three different levels. At the patent level, it changed the connectivity among patents A–C but made no structural changes at the IPC level nor at the USPC level. The number of U.S. patent classes (USPCs) assigned to a patent is usually much more than the number of IPC classes assigned. For example, six USPCs were assigned to patent A, but only two IPC classes were assigned to it.

We estimated the predictive effects of structural variation metrics using the citation-expanded patterns based on NCI patents. A baseline network was generated for each year with the citations of the top 100 most cited patents. Citation links made in the previous five years were retained for the baseline networks. The baseline networks were simplified to their minimum spanning trees. In addition to the three structural variation metrics, we included a few more independent variables such as the number of local existing links, the number of transformative links, and the number

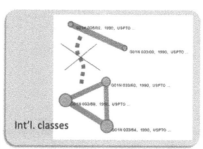

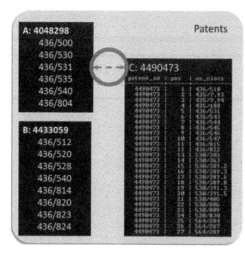

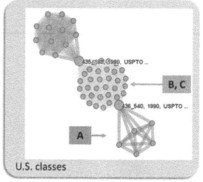

FIGURE 6.12 A patent may introduce structural changes at one level of granularity but not at other levels. Here patent 4676980 connected patents A and C, which were not connected at the patent level, but it didn't introduce any structural changes at the IPC level because it was assigned to the group of IPC classes that belong to the triangle. At the U.S. classification level, patents A, B, and C's classes are already connected.

of incremental links (see Table 6.4). The strongest predictor is the C_{KL} metric, with an IRR of 51.481. The second strongest predictor is the number of existing links.

Table 6.5 shows the results of an NB regression model of the expanded set of patents. The general pattern is similar. Two variables with statistically significant predictive effects are C_{KL} (IRR of 4.411) and local incremental links (1.010). To compare the ZINB and NB models, we used the Vuong nonnested hypothesis test to check whether the ZINB and NB models are indistinguishable. The Vuong test rejected the null hypothesis with a p-value of 0.0014 for −2.9859. The ZINB was superior.

Next, we added a few more independent variables to the ZINB and NB models, including local novel transformative links and local novel incremental links. Table 6.6 shows the results of the extended ZINB model. C_{KL} is still the strongest predictor, with an IRR of 22.2186. MCR is the second strongest, with an IRR of 1.0351.

The structural variation modeling identified U.S. Patent 6537746, with a score of 1.02, as the one with the highest C_{KL}. The patent was cited 22 times with an MCR of 39.75 and CL of 3.50. The patent generated 36 incremental links, that is, existing links, and 651 novel transformative links. Arnold Frances invented this patent, which was granted in 2003. Figure 6.13 shows the patent's various details.

Table 6.4   ZINB model of 2065 patents in the expanded NCI patents dataset

Theta = 0.3457

AIC(ZINB) = 8145.8	Count model coefficients			Zero-inflation model		
Variables	IRR	z	p	IRR	z	p
Intercept	0.000	−144.790	0.0000	0.1291	−4.141	0.0000
Local existing links	**1.021**	2.584	0.0098*	**0.098**	−2.220	0.0264*
Local transformative links	1.019	NA	NA	**14.083**	2.075	0.0380*
Local incremental links	1.003	1.903	0.0570	0.545	−1.728	0.0840
Modularity	1.020	1.272	0.2034	0.0116	−1.773	0.0763
Cluster linkage	0.027	NA	NA	0.000	−1.797	0.0724
Centrality divergence	**51.481**	4.749	0.0000*	629.302	1.892	0.0584

Boldfaced values are significant at $p < 0.05$.
*Significant at $p = 0.05$.

Table 6.5   NB model of the patent citations

Theta = 0.2788

AIC(NB) = 8176.7	Count model coefficients		
Variables	IRR	z	p
Intercept	2.668	21.875	0.000
Local existing links	1.016	1.865	0.062
Local transformative links	0.963	−0.681	0.496
Local incremental links	**1.010**	2.941	0.003*
Modularity	1.070	1.865	0.062
Cluster linkage	8.634	0.417	0.676
Centrality divergence	**4.411**	4.422	0.000*

Boldfaced values are significant at $p < 0.05$.
*Significant at $p = 0.05$.

Table 6.6   The ZINB model with more independent variables

Theta = 0.3504

AIC(ZINB) = 8146.052	Count model coefficients			Zero-inflation model		
Variables	IRR	z	p	IRR	z	p
Intercept	0.0001	−142.301	0.0000	0.1374	−4.146	0.0000
Local existing transformative links	1.1604	1.953	0.0508	0.6295	−0.290	0.7716
Local existing incremental links	**1.0216**	2.595	0.0095*	**0.0834**	−2.371	0.0177*
Local novel transformative links	0.9998	NA	NA	**20.1540**	2.103	0.0355*
Local novel incremental links	**1.0047**	2.619	0.0088*	0.4987	−1.771	0.0765
Modularity	**1.0351**	2.107	0.0351*	0.0061	−1.818	0.0690
Cluster linkage	0.1379	NA	NA	0.0000	−1.938	0.0526
Centrality divergence	**22.2186**	3.399	0.0007*	**287.4846**	2.189	0.0286*

Boldfaced values are significant at $p < 0.05$.
*Significant at $p = 0.05$.

Method for creating polynucleotide and polypeptide sequences

Frances Arnold et al

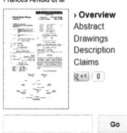

› Overview
Abstract
Drawings
Description
Claims

Patent number: 6537746
Filing date: Dec 4, 1998
Issue date: Mar 25, 2003
Application number: 09/205,448

The invention provides methods for evolving a polynucleotide toward acquisition of a desired property. Such methods entail incubating a population of parental polynucleotide variants under conditions to generate annealed polynucleotides comprising heteroduplexes. The heteroduplexes are then exposed to a cellular DNA repair system to convert the heteroduplexes to parental polynucleotide variants or recombined polynucleotide variants. The resulting polynucleotides are then screened or selected for the desired property.

Inventors: Frances Arnold, Zhixin Shao, Alexander Volkov
Original Assignee: Maxygen, Inc.
Primary Examiner: W. Gary Jones
Secondary Examiner: Janell E. Taylor
Attorney: Townsend and Townsend and Crew LLP
Current U.S. Classification: 435/6.12; 435/69.1; 435/91.1; 435/91.2; 435/320.1; 506/9; 536/23.7
International Classification: C12Q/168

View patent at USPTO
Search USPTO Assignment Database

FIGURE 6.13   U.S. Patent 6537746, which was identified with the highest C_{KL}.

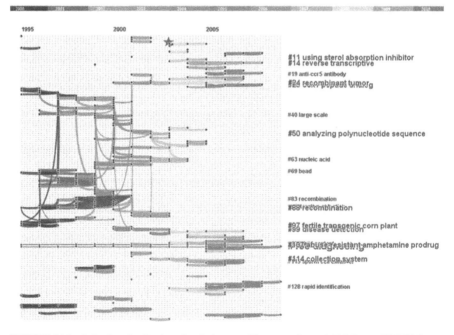

FIGURE 6.14   A timeline visualization of cocited patents. The star on the top is U.S. Patent 6537746. It contributed novel links connecting clusters from 83 through 88. (*See insert for color representation of the figure.*)

Figure 6.14 depicts the location of U.S. Patent 6537746, symbolized by a star on the top of the timeline visualization. Its issue date was 2003, between the labels of 2000 and 2005. The patent was ranked high in terms of C_{KL} because it contributed to the red novel links among clusters 83 through 88. The hybrid approach combines structural and statistical information so that we can trace along a chain of evidence to explain what makes a patent interesting and where we should direct our attention.

The structural variation theory addresses issues that are fundamental to science and technology. It is an explanatory and computational approach in that it explains how various mechanisms operate in a CAS and computationally synthesize multiple sources of information. It is also predictable in nature because one can recognize the state of a system in terms of the underlying mechanisms in action. These examples have illustrated the potential of the value-added approach to transform the workflow of scientific activity and shift the focus from access and retrieval to sensemaking and planning.

6.4.4 A Follow-Up Study

In July 2013, a Chinese journal, *Library Tribune*, published a study by two Chinese scholars. The two authors were Zhitao Yu and Xiaoqing Mu at the School of Foreign Languages of Shandong University of Technology in Zibo, China. They reviewed the structural variation model and explored a few examples of their own datasets.

Although the two authors previously published studies using CiteSpace, they did not seem familiar with the role of ZINB/NB models in our original study, published in the *Journal of the American Society for Information Science and Technology* (JASIST) in 2012.

The authors tested the structural variation model on their own datasets and commented that the model was particularly suitable for predicting novel developments in scientific literature, but the models did not perform as well when predicting future citations in areas central to an existing field. Central areas in an existing field are well known to researchers in the corresponding community. It is the peripheral areas that are of interest for novelty-seeking explorations. The primary concern of the structural variation approach is how to aid a search for turning points that may trigger critical transitions in the CAS.

The authors demonstrated how they could manipulate the citation patterns of a hypothetical paper and artificially boost structural variation scores of the paper. They commented that one can game the method to their own advantage.

Their experiments confirm the sensitivity of the structural variation metrics, which we believe is an advantage for a tool designed to detect early signs of critical transitions. Their concern excessively focused on how one may abuse the system. Rather than questioning if boundary spanning could be a valid mechanism for detecting novel connections and identifying early signs of change, their mindset seemed overlooked whether various technically possible actions should be also justified in a broader context. If an author could take all the trouble to manipulate his citing patterns in order to maximize his scores in a peer-reviewed system, he may well spend his time on something more constructive. Ironically, if one has all the knowledge and skills, but does not have a noble goal to achieve except to be nearsighted and aim at the ranking but not the substance of real contributions to scientific knowledge or the society, then they are probably clueless of what scientific inquiry is all about. Nevertheless, cybersecurity has become a widespread concern. Hackers tend to be technology focused and seek the challenges. There are people who are at an even lower moral level because their goal is to exploit the vulnerability of business systems. It is a practical issue. The authors' perspective may be valid in a rather peculiar perspective, but we would not be interested.

6.5 SUMMARY

We have set the structural variation theory in a broad context of detecting early signs of critical transitions in a CAS. A series of efforts in modeling the constructions of radical and high-impact patents have underscored the notion of recombinant search, which is consistent with the fitness landscape metaphor we introduced in Chapter 5. We have demonstrated an exploratory and computational procedure that can be applied to a number of domains of knowledge. In particular, the structural variation theory is implemented in terms of a recombinant search in a high-dimensional space of cocited references. Boundary-spanning mechanisms in such spaces are operationalized by detecting unprecedented between-cluster bridges or long-range novel connections. The extensibility of the approach is demonstrated in the example applications to scientific publications and patents.

The value of the approach is as follows:

* It provides intrinsic and instant metrics of scientific publications.
* It explains why a publication would be considered potentially transformative.
* It anticipates what if an area with missing links is filled.
* It can monitor the development of a body of knowledge as it grows.

In conclusion, the results are particularly encouraging. This approach offers not only the capability of making sense of why a specific connection is novel and valuable, but also insights into where challenges and opportunities may lie based on collective wisdom.

BIBLIOGRAPHY

Aksnes, D. W. (2003). Characteristics of highly cited papers. *Research Evaluation, 12*(3), 159–170.

Altshuller, G. (1999). *Innovation algorithm: TRIZ, systematic innovation and technical creativity* (1st ed.). Worcester, MA: Technical Innovation Center, Inc.

Barabasi, A. L., & Albert, R. (1999). Emergence of scaling in random networks. *Science, 286*(5439), 509–512.

Benedek, M., Fink, A., & Neubauer, A. C. (2006). Enhancement of ideational fluency by means of computer-based training. *Creativity Research Journal, 18*, 317–328.

Bilski v. Kappos, 130 S. Ct. 3218 (Supreme Court 2010). http://scholar.google.com/scholar_case?case=2277797231762274855&hl=en&as_sdt=4,60&sciodt=4,60&as_ylo=2009. Accessed on March 10, 2014.

Bornmann, L., & Daniel, H.-D. (2006). What do citation counts measure? A review of studies on citing behavior. *Journal of Documentation, 64*(1), 45–80.

Boyack, K. W., Klavans, R., Ingwersen, P., & Larsen, B. (2005). *Predicting the importance of current papers. Paper presented at the Proceedings of the 10th International Conference of the International Society for Scientometrics and Informetrics*, July 24–28, 2005, Stockholm, Sweden. Retrieved March 10, 2014, from https://cfwebprod.sandia.gov/cfdocs/CCIM/docs/kwb_rk_ISSI05b.pdf. Accessed on March 10, 2014.

Brody, T., & Harnad, S. (2005). *Earlier web usage statistics as predictors of later citation impact*. Retrieved July 15, 2013, from http://arxiv.org/ftp/cs/papers/0503/0503020.pdf. Accessed on March 10, 2014.

Bruderer, E., & Singh, J. (1996). Organization evolution, learning, and selection: A genetic-algorithm-based model. *Academy of Management Journal, 39*, 1322–1349.

Burt, R. S. (2004). Structural holes and good ideas. *American Journal of Sociology, 110*(2), 349–399.

Burton, R. E., & Kebler, R.W. (1960). The 'half-life' of some scientific and technical literatures. *American Documentation, 11*, 18–22.

Campbell, D. T. (1960). Blind variation and selective retentions in creative thought as in other knowledge processes. *Psychological Review, 67*(6), 380–400.

Chen, C. (2006). CiteSpace II: Detecting and visualizing emerging trends and transient patterns in scientific literature. *Journal of the American Society for Information Science and Technology, 57*(3), 359–377.

Chen, C. (2011). *Turning Points: The Nature of Creativity*. New York: Springer.

Chen, C., Chen, Y., Horowitz, M., Hou, H., Liu, Z., & Pellegrino, D. (2009). Towards an explanatory and computational theory of scientific discovery. *Journal of Informetrics, 3*(3), 191–209.

Chen, C., Ibekwe-SanJuan, F., & Hou, J. (2010). The structure and dynamics of co-citation clusters: A multiple-perspective co-citation analysis. *Journal of the American Society for Information Science and Technology, 61*(7), 1386–1409.

Chubin, D. E. (1994). Grants peer-review in theory and practice. *Evaluation Review, 18*(1), 20–30.

Chubin, D. E., & Hackett, E. J. (1990). *Paperless science: Peer review and U.S. science policy*. Albany, NY: State University of New York Press.

Collins, R. (1998). *The sociology of philosophies: A global theory of intellectual change*. Cambridge, MA: Harvard University Press.

Cuhls, K. (2001). Foresight with Delphi surveys in Japan. *Technology Analysis & Strategic Management, 13*(4), 555–569.

Cziko, G. A. (1998). From blind to creative: In defense of Donald Campbell's selectionist theory of human creativity. *Journal of Creative Behavior, 32*(3), 192–209.

van Dalen, H. P., & Kenkens, K. (2005). Signals in science: On the importance of signaling in gaining attention in science. *Scientometrics, 64*(2), 209–233.

Davis, M. S. (1971). That's interesting! Towards a phenomenology of sociology and a sociology of phenomenology. *Philosophy of the Social Sciences, 1*(2), 309–344.

Devezas, T. C. (2005). Evolutionary theory of technological change: State of the art and new approaches. *Technological Forecasting & Social Change, 72*, 1137–1152.

Dewett, T., & Denisi, A. S. (2004). Exploring scholarly reputation: It's more than just productivity. *Scientometrics, 60*(2), 249–272.

Fauconnier, G., & Turner, M. (1998). Conceptual integration networks. *Cognitive Science*, *22*(2), 133–187.

Fleming, L., & Bromiley, P. (2000). *A variable risk propensity model of technological risk taking*. Paper presented at the Applied Statistics Workshop, Harvard, MA. Retrieved June 15, 2010, from http://courses.gov.harvard.edu/gov3009/fall00/fleming.pdf. Accessed on March 10, 2014.

Friedrich, T. L., & Mumford, M. D. (2009). The effects of conflicting information on creative thought: A source of performance improvements or decrements? *Creativity Research Journal*, *21*(2–3), 265–281.

Garfield, E. (1955). Citation indexes for science: A new dimension in documentation through association of ideas. *Science*, *122*, 108–111.

Griliches, Z. (1990). Patent statistics as economic indicators: A survey. *Journal of Economic Literature*, *XXVIII*, 1661–1707.

Gruber, H. E. (1992). The evolving systems approach to creative work. In: D. B. Wallace & H. E. Gruber (Eds.), *Creative people at work: Twelve cognitive case studies* (pp. 3–24). Oxford, England: Oxford University Press.

Guilford, J. P. (1967). *The nature of human intelligence*. New York: McGraw-Hill.

Häyrynen, M. (2007). *Breakthrough research: Funding for high-risk research at the Academy of Finland*. Helsinki, Finland: The Academy of Finland.

Hennessey, B. A., & Amabile, T. M. (2010). Creativity. *Annual Review of Psychology*, *61*, 569–598.

Hettich, S., & Pazzani, M. J. (2006). *Mining for proposal reviewers: Lessons learned at the National Science Foundation*. Paper presented at the KDD'06, Philadelphia, PA.

Hilbe, J. M. (2011). *Negative binomial regression* (2nd ed.). New York: Cambridge University Press.

Hill, W. E. (1915). My wife and my mother-in-law. *Puck*, *16*, 11.

Hirsch, J. E. (2007). Does the h index have predictive power? *Proceedings of the National Academy of Sciences*, *104*(49), 19193–19198.

Hudson, L. (1966). *Contrary imaginations: A psychological study of the English schoolboy*. New York: Schocken.

Hunter, S. T., Bedell-Avers, K. E., Hunsicker, C. M., Mumford, M. D., & Ligon, G. S. (2008). Applying multiple knowledge structures in creative thought: Effects on idea generation and problem-solving. *Creativity Research Journal*, *20*(2), 137–154.

Kostoff, R. (2007). The difference between highly and poorly cited medical articles in the journal *Lancet*. *Scientometrics*, *72*(3), 513–520.

Koza, J. R., Al-Sakran, S. H., & Jones L. W. (2005, June 25–29). Automated re-invention of six patented optical lens systems using genetic programming. *Genetic and Evolutionary Computation Conference (GECCO) '05* (pp. 1953–1960), Washington, DC.

Koza, J. R., Keane, M. A., & Streeter, M. J. (2003, February). Evolving inventions. *Scientific American*, *288*(2), 52–59.

Kuhn, T. S. (1977). *The essential tension: Selected studies in scientific tradition and change*. Chicago/London: University of Chicago Press.

Kurtz, M. J., Eichhorn, G., Accomazzi, A., Grant, C., Demleitner, M., Henneken, E., et al. (2005). The effect of use and access on citations. *Information Processing & Management*, *41*(6), 1395–1402.

Lambert, D. (1992). Zero-inflated Poisson regression, with an application to defects in manufacturing. *Technometrics, 34*, 1–14.

Lauronen, E., Veijola, J., Isohanni, I., Jones, P. B., Nieminen, P., & Isohanni, M. (2004). Links between creativity and mental disorder. *Psychiatry, 67*(1), 81–98.

Lavery, D. (1993). Creative work: On the method of Howard Gruber. *Journal of Humanistic Psychology, 33*(2), 101–121.

Levinthal, D. (1997). Adaptation on rugged landscapes. *Management Science, 43*, 934–950.

Levitt, J., & Thelwall, M. (2008). Patterns of annual citation of highly cited articles and the prediction of their citation ranking: A comparison across subjects. *Scientometrics, 77*(1), 41–60.

Lipinski, C., & Hopkins, A. (2004). Navigating chemical space for biology and medicine. [Article]. *Nature, 432*(7019), 855–861.

Luxburg, U. v. (2006). *A tutorial on spectral clustering.* Retrieved March 15, 2012, from http://www.kyb.mpg.de/fileadmin/user_upload/files/publications/attachments/Luxburg07_tutorial_4488%5b0%5d.pdf. Accessed on March 10, 2014.

Maddux, W. W., Adam, H., & Galinsky, A. D. (2010). When in Rome … Learn why the Romans do what they do: How multicultural learning experiences facilitate creativity. *Personality and Social Psychology Bulletin, 36*(6), 731–741.

Martin, B. R. (2010). The origins of the concept of 'foresight' in science and technology: An insider's perspective. *Technological Forecasting and Social Change, 77*(9), 1438–1447.

McPherson, J., & Ranger-Moore, J. (1991). Evolution on a dancing landscape: Organizations and networks in dynamic Blau space. *Social Forces, 70*, 19–42.

Merton, R. K. (1968). The Mathew Effect in science. *Science, 159*(3810), 56–63.

Miles, I. (2010). The development of technology foresight: A review. *Technological Forecasting and Social Change, 77*(9), 1448–1456.

Nemet, G. F., & Johnson, E. (2012). Do important inventions benefit from knowledge originating in other technological domains? *Research Policy, 41*, 190–200.

Newell, A., Shaw, J. C., & Simon, H. A. (1958). Elements of a theory of human problem-solving. *Psychological Review, 65*(3), 151–166.

Newman, M. E. J. (2006). Modularity and community structure in networks. *Proceedings of the National Academy of Sciences of the United States of America, 103*(23), 8577–8582.

Nocaj, A., & Brandes, U. (2012). Computing Voronoi Treemaps: Faster, simpler, and resolution-independent. *Computer Graphics Forum, 31*(3), 855–864.

Persson, O. (2010). Are highly cited papers more international? *Scientometrics, 83*(2), 397–401.

Price, D. D. (1965). Networks of scientific papers. *Science, 149*, 510–515.

Rothenberg, A. (1987). Einstein, Bohr, and creative-thinking in science. *History of Science, 25*(68), 147–166.

Rothenberg, A. (1996). The Janusian process in scientific creativity. *Creativity Research Journal, 9*(2–3), 207–231.

Scheffer, M., Bascompte, J., Brock, W. A., Brovkin, V., Carpenter, S. R., Dakos, V., et al. (2009). Early-warning signals for critical transitions. *Nature, 461*(7260), 53–59.

Schilling, M. A., & Green, & E. (2011). Recombinant search and breakthrough idea generation: An analysis of high impact papers in the social sciences. *Research Policy, 40*, 1321–1331.

Scott, G. M., Leritz, L. E., & Mumford, M. D. (2004). The effectiveness of creative training: A quantitative review. *Creativity Research Journal, 16*, 361–388.

Simonton, D. K. (1999). *Origins of genius: Darwinian perspectives on creativity.* New York: Oxford University Press.

Sorenson, O. (2000). The effect of population level learning on market entry: The American automobile industry. *Social Science Research, 29*, 307–326.

Swanson, D. R. (1986a). Fish oil, Raynaud's syndrome, and undiscovered public knowledge. *Perspectives in Biology and Medicine, 30*(1), 7–18.

Swanson, D. R. (1986b). Undiscovered public knowledge. *Library Quarterly, 56*(2), 103–118.

The Delphian Society. (1913). *The world's progress, Part III.* Hammond, IN: W. B. Conkey Company.

Tichy, G. (2004). The over-optimism among experts in assessment and foresight. *Technological Forecasting and Social Change, 71*(4), 341–363.

Upham, S. P., Rosenkopf, L., & Ungar, L. H. (2010). Positioning knowledge: Schools of thought and new knowledge creation. *Scientometrics, 83*, 555–581.

Waddell, C. (1998, March). Creativity and mental illness: Is there a link?. *Canadian Journal of Psychiatry, 43*(2):166–172.

Walters, G. D. (2006). Predicting subsequent citations to articles published in twelve crime-psychology journals: Author impact versus journal impact. *Scientometrics, 69*(3), 499–510.

Watts, D. J., & Strogatz, S. H. (1998). Collective dynamics of 'small-world' networks. *Nature, 393*(6684), 440–442.

Weeber, M. (2003). Advances in literature-based discovery. *Journal of the American Society for Information Science and Technology, 54*(10), 913–925.

Yu, Z. Y., & Mu, X. (2013). Chen's indicators for scientometrics and their application. *Library Tribune, 33*(4), 32–41.

Zeileis, A., Kleiber, C., & Jackman, S. (2011). *Regression models for count data in R.* Retrieved July 15, 2013, from http://cran.r-project.org/web/packages/pscl/vignettes/countreg.pdf. Accessed on March 10, 2014.

Zhao, H., & Jiang, G. (1985). Shifting of world's scientific center and scientists' social ages. *Scientometrics, 8*(1–2), 59–80.

CHAPTER 7

Gap Analytics

On September 27, 2013, the Office of the Director of National Intelligence of the Intelligence Advanced Research Projects Activity (IARPA) issued a request for information (RFI).[1] The RFI is seeking information on methods for analyzing and forecasting trends and milestones in science and technology.

The RFI provides a list of illustrative examples of the types of questions they would like to address:

1. "Which companies lead the world in organic light-emitting diode (OLED) manufacturing?"
2. "What is the probability of a 10 cm carbon nanotube being fabricated before December 31, 2014?"
3. "Among abstracts accepted for the 2015 International Conference on Machine Learning (ICML) conference, will the number containing the term 'deep learning' exceed the number containing the term 'support vector machine(s)'?"
4. "How many unique assignees will have at least two USPTO patent applications published using the term 'Type III Secretion System' in their title/abstract/background/claims between October 1, 2013 and September 30, 2014?"
5. "By December 31, 2017, how many FDA-approved products will be based on RNA interference?"
6. "Will there be reported shortages of technetium-99 m in the United States in 2015?"

Let's pick question 2 and consider how we may address this type of question.

The key to addressing this question is to be able to estimate the maximum length of a carbon nanotube at some point in the future. Similarly, questions may limit the scope to a particular country, such as the United States or China, impose an upper limit of the cost, or specify other types of constraints. In its simplest form, a good strategy would be to first find out the state of the art worldwide. Next, we would find out the timeline of how fast advances have been made in the past 5 or 10 years in

[1] http://www.iarpa.gov/RFI/rfi_sti.html

The Fitness of Information: Quantitative Assessments of Critical Evidence, First Edition. Chaomei Chen.
© 2014 John Wiley & Sons, Inc. Published 2014 by John Wiley & Sons, Inc.

making longer carbon nanotubes. Then, we would need to identify existing challenges that prevent anyone from making a 10 cm carbon nanotube and what techniques are currently available to address individual challenges. Finally, if there is no known solution to an existing problem, how likely is it to find a feasible solution within the specified time frame? To answer the seemingly simple question 2, it appears that we need to be able to answer a series of questions. A capable method, therefore, will likely be an integration of multiple methods.

When I tried to type the query "carbon nanotube length record" on Google, someone else had already searched with the exact query. Perhaps many people are considering how to answer question 2. The first link returned by Google was dated July 10, 2013. It reported that Tsinghua University in China was able to grow a carbon nanotube over 50 cm long (Zhang et al., 2013), which seems that its length has already met the requirement in question 2. We have two options. One is to answer question 2 with a probability of 1 that a 10 cm carbon nanotube can be made by the end of 2014 because there is at least one piece of evidence to support the prediction. The other option is to be more cautious and do more research before answering the question. Is the record-length carbon nanotube reproducible? Is it reproducible by researchers in other countries? What we have learned from OPERA's over-the-speed-of-light finding at the beginning of this book is that we need to ask if we are seeing the tip of an iceberg.

Searching for the world record of the length of a carbon nanotube through unstructured text is not straightforward. Searching for the world record within a specific time period is even harder. Question 2 has revealed several profound challenges. Similar problems may be found in electronic health-care records and how clinical trials specify inclusion and exclusion criteria in freely flowing natural languages. The need to assess the state of the art and track the gap between where we are and where we want to be is a common core in portfolio analyses.

7.1 PORTFOLIO ANALYSIS AND RISK ASSESSMENT

Portfolio analysis has a critical role in strategic planning, risk assessment, policy making, and performance evaluation. It is concerned with a broad spectrum of scientific and technological domains. The primary goal of a portfolio analysis is to assess the performance of a unit of interest, such as an individual, an organization, or a discipline, and identify its strengths and weaknesses regarding a baseline such that strategic adjustments can be made accordingly. Obtaining a holistic picture of the unit of analysis as a complex adaptive system is therefore of profound significance. Methodologies of portfolio analysis can be applied to a wide range of application domains, including gap analysis, situation awareness, competitive intelligence, and research evaluation and assessments.

An important characteristic of a complex adaptive system is that the whole is usually greater than the sum of its parts. In addition to studying individual components of such a system, it is essential to study how individual components are interrelated and how such interrelationships change over time in response to external events and internal perturbations. To be able to cover the structure and dynamics both at the component level and at the system level, analysts face a tremendous challenge to associate patterns identified at one level with patterns identified at another level. In this chapter, we will focus on how this issue can be addressed in the context of portfolio analysis of publications produced by a unit of interest, including individual scientists, university colleges, research institutes, funding organizations, and scientific fields.

The notion of global science maps and local science maps has been seen in the literature, especially in information science and information visualization. Global science maps, for example, focus on depicting interrelationships of disciplines, whereas local science maps often focus on a specific field of study or a specialty. These existing approaches to the global and local science mapping are limited in terms of the types of organizing frameworks that can be offered to accommodate a portfolio analysis. A typical use of a global science map is to provide a base map over which a layer of additional information, or an overlay, can be superimposed. While existing solutions such as interactive overlays can provide insightful findings of research groups, many potentially significant analytical tasks are not readily supported. For example, each instance of citation in a publication involves a source and a target. The source is the article that initiates the citation, whereas the target is the reference that is being cited. To our knowledge, no global map overlays explicitly depict sources and targets of citations simultaneously.

We will first review an example of a portfolio evaluation in the context of a funding program. Then, we will introduce interactive overlays on journal-based global maps of science. Furthermore, we will introduce interactive overlays on a dual-map design and follow by a few examples to demonstrate how this method can be applied to the analysis of a portfolio.

7.1.1   Portfolios of Grant Proposals

The NSF CISE/SBE Advisory Committee formed a Subcommittee on Discovery in Research Portfolios between 2009 and 2010. The subcommittee was charged with identifying and demonstrating techniques and tools that can facilitate the assessment and evaluation of grant proposals and award portfolios. The subcommittee was asked to identify tools and approaches that are most effective in deriving knowledge from a diverse range of data. The tools should enable program directors visualize, interact, and understand the knowledge derived from the data. Subcommittee members were asked to apply their research tools to structure, analyze, visualize, and interact with datasets provided by the NSF.

Grant proposals submitted to the NSF consist of a number of components, including a cover page, a one-page project summary, a project description up to 15 pages, a list of references, 2-page biographies of investigators, and budget information. The abstracts of awarded projects are publicly available on NSF's website. A set of proposals were selected and made available to the members of the subcommittee for a limited period of time, but reviews of proposals were not accessible. All the results discussed in this book regarding the proposal dataset have been approved by a specific clearance procedure, which was in place to safeguard the privacy and security of the proposal dataset.

We will focus on two types of questions. At the individual proposal level, the main questions are: What is a proposal about in a nutshell? How does one proposal differ from other proposals in terms of their nutshell representations? At the portfolio level, the questions focus on characteristics of a group of proposals. What are the computational indicators that may differentiate awarded and declined proposals? What are the indicators that may identify transformative proposals in a portfolio?

Identifying the Core Information of a Proposal

We make no assumption about the structure or content of text documents. Proposals are processed as unstructured text. We expect that the amount of core information in a proposal is likely to be more than a one-page summary but shorter than a 15-page full-length proposal. If this is true, then it would make sense to reduce the text in a proposal to its core information and still preserve the essence of the full proposal. It would be also reasonable to expect that the shortened representation is probably more coherent than the original full-length document.

One way to extract the core information from a full-length proposal is as follows. First, divide the full-length project description of the proposal into a series of passages of text (known as segments) so that each passage corresponds to an underlying theme or topic. Next, construct a network of these passages based on their similarities. Finally, select passages of high centrality scores in the network to represent the core information of the proposal.

Figure 7.1 illustrates this process. The full-text document was divided into a sequence of segments. The plot in the middle depicts similarities between adjacent segments using a sliding window. A network of segments was generated based on intersegment similarities. The segments in darker shade played a central role in the network. Both of them can be selected to represent the essence of the full proposal.

It is possible to divide text into segments based on the degree of cohesiveness. A passage of text can be partitioned in many ways when the number and sizes of the segments are considered. Given a particular partition, the internal cohesiveness of each segment can be measured by techniques such as latent semantic indexing. The optimal partition would be the one that maximizes cohesiveness of all individual segments. An alternative way to optimize the partition is to ensure that the internal cohesiveness of text within a segment is higher than between segments. The process is known as *text*

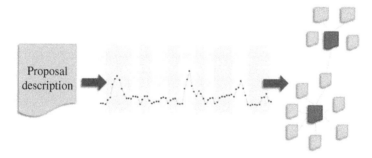

FIGURE 7.1   The procedure for identifying core passages of a full-length document.

segmentation. The basic assumption is that a text document usually represents a series of subtopics and it is possible to detect the boundary of a subtopic based on the change of text similarities. Marti Hearst developed a flexible text segmentation algorithm, which can be used for this purpose. Optimizing the parameters for the algorithm is crucial, but currently computationally optimizing the configuration does not appear to be available. Using interactive visualization is a practical solution. Users could explore different configurations and select a good one.

Hearst's text segmentation algorithm detects a shift of topic based on lexical differences of a set of *n*-grams of text. The value of *n* ranges from 10 to 200 units. We implemented this method and provided an interactive visualization of lexical differences to help users to find the optimal parameter combinations. *Window size* and *step size* are two parameters crucial in configuring the text segmentation process. Window size is the number of tokens, mainly terms excluding stop words, in a sequence of tokens, and step size is the number of sequences of token in a block that are used for block–block similarity comparison. The similarity between two blocks is measured by a normalized inner product: given two blocks, b_1 and b_2, each with *window-size* token sequence, where $b_1 = \{\text{token-sequence}_{i-k}, ..., \text{token-sequence}_i\}$ and $b_2 = \{\text{token-sequence}_{i+1}, ..., \text{token-sequence}_{i+k+1}\}$. Our test with a small number of texts found that window size = 100 and step size = 20 yielded the best results in narrative texts similar to the NSF proposals.

The next step is to select the most representative segments as the core information of a proposal. Once text segments are identified, the similarity between any two segments can be calculated, including vector space models, latent semantic indexing, probabilistic models, or more recent topic models. A network of segments can be constructed based on intersegment similarities. Choosing the most representative block of text is equivalent to choosing the segments with high centrality measures in the network. A central topic is expected to be highly connected with other topics in the same proposal. Metrics such as PageRank are able to rank segments in the order of such centrality. One may choose one or multiple top-ranked segments to represent the proposal for any subsequent text analysis, clustering, or visualization. Our study evaluated candidate ranking metrics against our own proposals and found that segments selected by PageRank are more meaningful than other options.

Information Extraction

Information extraction is process unstructured text and select words that are most relevant to our interest. Natural language processing (NLP) techniques are commonly used in information extraction. It is generally believed that noun phrases are more meaningful and interpretable than single words. In the extraction step, noun phrases are extracted from core segments identified in a proposal.

Noun phrases consist of multiple words with a noun as the final word (known as the head noun). For example, the head noun in the noun phrase *supermassive black hole* is black hole. A *term* may be a single word or multiple words, but does not necessarily contain nouns. Sometimes terms are also referred to as *n*-grams, with *n* as the number of component words. Noun phrases are considered in general better lexical units to represent concepts than words and terms because noun phrases tend to be more meaningful and self-contained.

In order to identify noun phrases in unstructured text, the first step is to tag the type of each word in a text passage, including nouns, verbs, and adjectives. This step is called part-of-speech (POS) tagging. NLP tools are available to perform POS tagging and process tagged text, for example, the GATE system and the Stanford NLP Toolkit. Since NLP tools tend to be built with particular training text, the quality of tagging varies from target datasets. However, we found that using regular expression is the most flexible, customizable, and extensible approach. We experimented with several types of noun phrases in terms of the number of nouns because a general form of a noun phrase is word–word–word–noun and it is possible that the words are also nouns themselves, for example, word–word–noun–noun as in *rapidly increased climate change*. We allow users to filter noun phrases in terms of the number of nouns in a phrase. We tested analytical and statistical results across noun phrases with different word counts.

Detecting Hot Topics

Hot topics are defined in terms of the frequency of noun phrases found in project descriptions, project summaries, or other sources of text. Generally speaking, high-frequency noun phrases are regarded as indicators of a possible hot topic. The most valuable information about a hot topic is since when it becomes hot and how long it will last.

Burst detection determines whether the frequency of an observed event is substantially increased over time with respect to other events of the same type. The types of events are generic, including the appearance of a keyword in newspapers over a period of 12 months and the citations to a particular reference in papers published in the past 10 years. The data mining and knowledge discovery community has developed several burst detection algorithms. There are many techniques for detecting the emergence of a hot topic, notably Kleinberg's burst detection algorithm.

Two temporal properties of the burst of a noun phrase are the waiting time to burst and the duration of burst. The waiting time to burst is how much time has elapsed between the initial appearance of a noun phrase in a set of proposals and when a burst is detected by the algorithm. The duration of a burst is the time elapses between the beginning of the burst until either the burst is over or the end of the time frame of the analysis is reached. These properties are used in a survival analysis to differentiate awarded and declined proposals. Since these properties are domain independent, this method is applicable to a wide range of domains.

Identifying Potentially Transformative Proposals

We envisage that transformative research should be computationally detectable along two dimensions: *synthesis distance* and *structural divergence*. The synthesis distance characterizes a particular scientific contribution in terms of the conceptual distance between component topics it synthesizes and integrates. It is harder to conceive a long-range synthesis than a short-range one, but a synthesis over a long distance is likely to have a higher level of novelty. The structural divergence measures the extent to which a particular scientific contribution departs from the state of the art. As illustrated in Figure 7.2, groundbreaking ideas are likely to have a distant synthesis distance and a large structural divergence.

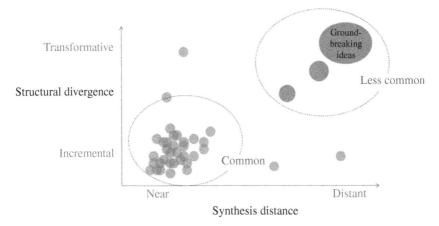

FIGURE 7.2   An illustration of the distribution of transformative research along two dimensions.

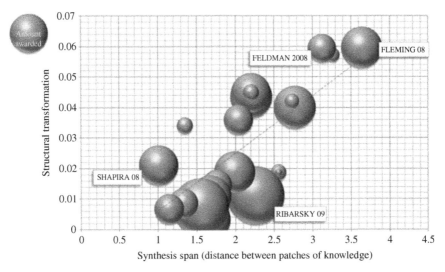

FIGURE 7.3   Transformative potentials of awards. Source: Publicly available NSF award abstracts of the SciSIP program.

Figure 7.3 shows the results of an analysis of proposals awarded in the NSF SciSIP program. Each proposal was represented by its publicly available abstract. The position of an award is determined by its synthesis distance and structural divergence scores. The size of each award represents the amount awarded. Further details of the four awards annotated in Figure 7.3 are listed in Table 7.1, including investigators' names, the year of the award, and the title of the project.

We also analyzed 200 proposals (100 awarded and 100 declined) randomly sampled from 7345 proposals of a different NSF program. The core information of each proposal was represented by three core segments with the highest PageRank scores.

Table 7.1   Awards labeled in Figure 7.3

Investigator	Year	Title
Lee Fleming	2008	DAT: Creating a Patent Collaboration Network Database to Examine the Social Production of Knowledge
Feldman Maryann	2008	State Science Policies: Modeling Their Origins Nature Fit and Effects on Local Universities
Martin Ribarsky	2009	DAT: A Visual Analytics Approach to Science and Innovation Policy
Philip Shapira	2008	MOD Measurement and Analysis of Highly Creative Research in the United States and Europe

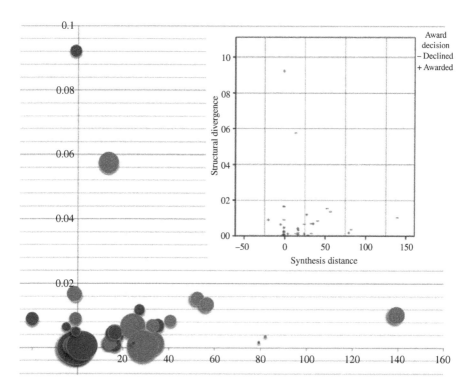

FIGURE 7.4   Transformative potentials of proposals (+: awarded; –: declined). Size=the amount requested. Data Source: 200 randomly sampled proposals.

Noun phrases of one to four nouns were extracted by CiteSpace to estimate the transformative potential properties for each proposal. A total of 141 proposals (70.5%) were found to have positive readings on the chart (see Figure 7.4).

A lesson learned from the analysis of grant proposals was that a broader context is needed to characterize not only the position of an idea at a particular time, but also the trajectory of where it has been. In the next section, we will turn to the design of interactive overlays on a global map of scientific literature and demonstrate how to analyze a portfolio in terms of the movement of an underlying organization.

7.2 INTERACTIVE OVERLAYS

A commonly used strategy in cartography is to superimpose a thematic layer over a geographic map. For instance, store locators typically display the locations of their stores on a road map or a satellite map. The base map usually refers to the generic geographic map, whereas a thematic overlay refers to a layer of a special type of geographically distributed information. Multiple thematic overlays can be used to display related or nonrelated information. For example, tourists in a city may find it useful to look at an overlay of restaurants and an overlay of ATMs simultaneously. A major advantage of the overlay strategy is that we are familiar with how a map should be used, which tends to reduce the perceived complexity substantially. Sometimes this advantage may become crucial because we tend to be put off by something that appears too complicated. If we can bear with the initial impression of the complexity, then we are more likely to stay and start to learn something new.

7.2.1 Single-Map Overlays

Interactive overlays have been used with a variety of nongeographical base maps. The advantage of having a well-understood base map is still valid. Once we have learned how information is organized in a nongeographical map, we are able to focus on what information in an overlay brings to us and how it fits with our existing understanding of the base map.

Researchers have been developing ways to generate global science maps and use them to facilitate the analysis of issues concerning interrelated disciplines and the interdisciplinarity of a research program. Ismael Rafols, Loet Leydesdorff, and Alan Porter have been studying interdisciplinary research, especially topics that have profound societal challenges such as climate change and the diabetes pandemic. Addressing such societal challenges requires communications and incorporations of different bodies of knowledge, both from disparate parts of academia and from social stakeholders.

Interdisciplinary research involves a great deal of cognitive diversity. Rafols, Porter, and Leydesdorff (2010) developed a method based on overlaying science maps to study interdisciplinary research. The overlay method has two steps: (1) creating a global map of science as the base map and (2) superimposing a layer of a set of publications, for example, from a given institution or on a chosen topic. They have developed a toolkit and made it available for everyone to use.[2] A collection of interactive science overlay maps are also available on the Web.[3] With these interactive maps, one can explore how publications of an organization are distributed in various

[2] http://www.leydesdorff.net/overlaytoolkit
[3] http://idr.gatech.edu/maps.php

disciplines. These maps have been used to study the interdisciplinarity of research (Porter & Rafols, 2009), to compare research outputs of universities and large corporations (Rafols et al., 2010), and to trace the diffusion of research topics over science (Leydesdorff & Rafols, 2011).

The interactive overlay maps created by Rafols, Porter, and Leydesdorff are based on a single base map of scientific journals. Their base maps represent a network of journals based on their citing patterns or based on their cited patterns. In each view, there is only one representation of the global structure of scientific knowledge. This design has some limitations. The major limitation is the separation of citing and cited aspects of a citation. Each overlay can only represent one aspect of a citation, which does not seem to be feasible to address questions such as the following:

- Given a distribution of a set of articles on a base map of citing journals, what is their distribution on the *counterpart* cited journal base map?
- Which journals were cited by articles published in a particular journal?
- How do two organizations differ in terms of where they publish *and* what they cite?

In order to address these questions and other questions central to portfolio analysis, we have introduced a novel design that uses dual-map overlays. Dual-map overlays can address all the questions that single-map overlays can answer. In addition, dual-map overlays provide insights that would not be possible with a single-map overlay.

7.2.2   Dual-Map Overlays

Many citation-based maps are designed to show either the sources or the targets of citations in a single display but not both. A primary concern is that representing a mixture of citing and cited patterns simultaneously may considerably increase the complexity for users. HistCite, for example, represents the citation relationship of articles as a graph. A node in a graph cites nodes in previous years, whereas the node itself may be cited by nodes in the following year. The structure of such a graph is determined by the set of articles being considered. Understanding the structure of one dataset does not help us much to understand the structure of a different dataset. Furthermore, representing all the citation links in such graphs may suffer from a reduced clarity and make it harder for us to accomplish our analytic tasks. Although it is conceivable that a combined structure may be desirable in situations such as a heated debate, researchers are in general more concerned with differentiating various arguments before considering how to combine them.

The Butterfly system designed in the mid-1990s by Jock Mackinlay and his colleagues at Xerox displayed both citing and cited information in the same view, but each view was centered around a single article instead of presenting a holistic view of a collection of articles or journals (Mackinlay, Rao, & Card, 1995). Eugene Garfield's HistCite depicted direct citations in the literature. However, as the number of citations increase, the network tends to become cluttered, which is a common problem to network representations (Garfield, Pudovkin, & Istomin, 2003; Lucio-Arias & Leydesdorff, 2008).

We will introduce a novel design that uses dual-map interactive overlays to reveal additional insights into the structure and dynamics of citation patterns. The dual-map overlay design has several advantages over a single-map overlay. First, it represents the entirety of a citation instance. One can see where a citation originates and where it leads to in a single noninterrupted view. Second, it is easier to compare patterns of citation links than patterns of distributed dots. Third, it provides a framework for a new type of portfolio analysis with reference to a variety of evidence that characterize the movement of multiple populations across an adaptive landscape of scientific knowledge. Fourth, it opens up more research questions that can be addressed in new ways of analysis. For example, it becomes possible to study the interdisciplinarity at both source and target sides. It becomes possible to track the movements of scientific frontiers in terms of their footprints in both base maps.

The new design resembles the metaphor of fitness landscapes (Wright, 1932) in many ways. We can naturally introduce the notion of a trajectory of a collection of scientific publications, a set of patentable ideas, or a series of decisions made by the Supreme Court. A trajectory of a set of publications can be computed at the level of journals or at a disciplinary level.

The entire set of journals can be partitioned into groupings of journals, either by citing and cited patterns. These groupings may represent disciplines or research fields, depending on the granularity chosen for the partition process.

We used Blondel et al.'s algorithm (Blondel, Guillaume, Lambiotte, & Lefebvre, 2008) to partition the networks of journals in examples described in this chapter. We refer to such clusters as Blondel clusters. Ludo Waltman and Nees Jan van Eck recently announced an algorithm, called the smart local moving (SLM) algorithm, for community detection in large networks. The algorithm maximizes the modularity of a network to find the best partition. According to Waltman and van Eck, the SLM algorithm has been successfully applied to networks with tens of millions of nodes and hundreds of millions of edges (Waltman & van Eck, 2013). It would be a strong candidate for this type of algorithms.

A trajectory of an agent is the path that the agent has traversed in a space of latent variables, which could be a high-dimensional space as well as a 2- or 3-dimensional space. The agent could be an individual, a subpopulation (e.g., an organization), or the entire population. A trajectory *path* in a dual-map visualization is a function defined on a subset of the space × time domain to a subset of the fitness surface: $\text{path}(x(t)) = (x(t), \text{fitness}(x(t)))$. At a given time t, $x(t)$ is the position of the agent on the base map, thus $\text{fitness}(x(t))$ represents the fitness value of the agent at time t. The point $(x(t), \text{fitness}(x(t)))$ is a location on the fitness landscape that is occupied by the agent at time t.

The unit of time t can be year, month, hour, or minute. In our case, the unit is either year or month because of the resolution of the data. Each overlay layer represents a set of publication. As we will see shortly, such a set defines a portfolio of publications, depending on how the dataset is constructed. An overlay dataset D consists of n articles a_i, $i = 1, 2, \ldots, n$. Each article a_i appears in a journal b_j. The positions of the journal on the two base maps are $\text{citing}(b_j)$ and $\text{cited}(b_j)$, respectively. All the articles in D that are published in year t form $D(t)$, which is a subset of D. The positions of $D(t)$ on the base maps are the weight centers of $\{\text{citing}(b_j)\}$ and $\{\text{cited}(b_j)\}$. The use of a weight center

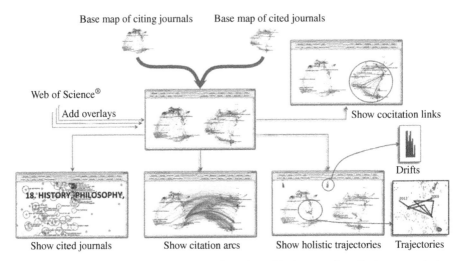

FIGURE 7.5 An overview illustrates the construction and use of dual-map overlays. Citation arcs, cocitation links, and trajectories over time facilitate the study of multiple sets of publications at an interdisciplinary level, an organizational level, and the individual publication level.

is found in the literature in aggregating the information from multiple points, for example, the computation of a Barycenter (Jin & Rousseau, 2001).

The construction of a dual-map base shares the initial steps of interactive overlay maps but differs in later steps. Once the coordinates are available for both citing and cited matrices of journals, a dual-map overlay can be constructed. It is not necessary to have cluster information, but additional functions are possible if cluster information is available. We assume that at least one set of clusters are available for each matrix. In this example, clusters are obtained by applying the Blondel clustering algorithm. Figure 7.5 shows an overview of the new method for a dual-map based portfolio analysis of scientific publications. The method is extensible to other types of global base maps, but here we will limit our descriptions to base maps generated from JCR journals. Details of the base map generation can be found in (Leydesdorff, Rafols, & Chen, 2013).

In the context of scientific publications, each member of a portfolio is a source article, also known as a citer, or a citing article. The journal in which a source article is published is called a citing journal. A reference cited by a source article is called a cited article, or a target article of an instance of citation, which may or may not be a source article in its own right. The journal in which a reference is published is called a cited journal. References cited by the same source article are called cocited references. The publication date of a source article can be identified either as the year only or the year and the month of the publication. The publication date of a cited reference is the year in which it is published. A portfolio represents the output of a research unit, whereas the references it cites as a whole represent the knowledge base on which the research activity is built.

A portfolio is a set of publications of interest. A portfolio can be constructed in a wide variety of ways. Portfolios are commonly defined by a common basis of authorship.

For example, an individual scientist's portfolio consists of all the publications authored or coauthored by the same author. A university's portfolio consists of all the publications by members of the same university. A country's portfolio similarly consists of all the publications by authors in the same country. On the other hand, a set of publications can be collected such that they represent the output of scientists in a particular field of study as a whole. Similarity, one can compile a dataset that represents the activity of a discipline. We can also construct a portfolio with a seed publication and group all the publications related to the seed in one way or another. These arguments can be made for constructing portfolios of patents invented by NCI employees, or on carbon nanotubes or telescopes, and grant proposals funded by the same program.

Given a base map of citing journals and a base map of cited journals, the two base maps are presented in the same user interface. The source of an overlay is a set of bibliographic records retrieved from the Web of Science and stored in a file directory on a computer where the software runs. For each overlay, the user may designate a specific color to distinguish citation arcs that belong to different overlays (see the lower middle part of Figure 7.5). The color chosen by the user will be also used for the trajectories of the overlay. Each trajectory is depicted in a bar chart that shows the pace lengths of the moves made by the trajectory and a trajectory plotted on the two-dimensional base maps (shown in the lower right part of Figure 7.5). The starting time and the ending time of each trajectory are marked. Each segment of a trajectory points from the end with a darker color to the end with a brighter color. The circled area in the upper right part of Figure 7.5 shows cocitation links from an overlay.

Given an overlay, journals involved in the citing and cited base maps are marked with circles (as shown in the lower left part of Figure 7.5). All the journals on a base map are assigned to clusters obtained by the Blondel algorithm (Leydesdorff et al., 2013). Major clusters are labeled by terms chosen from the titles of journals in corresponding clusters. The label terms are selected by a log-likelihood ratio test algorithm implemented in CiteSpace (Chen, 2006; Chen, Ibekwe-SanJuan, & Hou, 2010). For example, the cluster in the lower left part of Figure 7.5 is labeled by terms such as history and philosophy.

Multiple overlays can be superimposed onto the dual-map base one by one. An existing overlay can be removed. The user may use a number of controls such as buttons and sliders to select various types of information to be displayed. The following examples used the same base maps with the Blondel clustering configuration. Each side of the dual-map base depicts over 10,000 journals.

Figure 7.6 shows an annotated user interface. It shows both the citing and cited base maps side by side. The citing base map of 10,330 citing journals is on the left and the cited one of 10,253 cited journals is on the right. Each dot is a journal. Its color denotes its Blondel cluster membership. Various controls are available, for example, switching between Blondel clusters and VOSviewer's clusters, switching the unit of time between yearly and monthly (YR≥MTH and MTH≥YR), and switching between the calculation of trajectories at the cluster or journal level (C≥J and J≥C). The link style at the upper right controls the style of citation links. Our current design provides two types of styles, namely, curves and arcs. The arc style depicts a citation link as a parabolic arc. The curve style depicts a citation link as a spline curve running

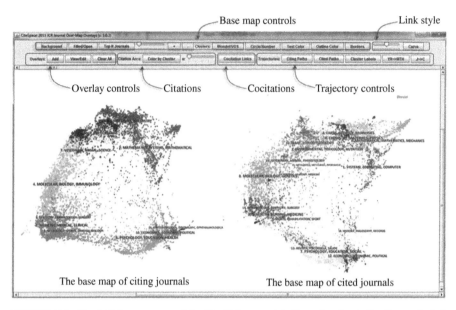

FIGURE 7.6   The initial appearance of the Dual-Map user interface, showing both citing and cited journal base maps simultaneously. The base map of 10,330 citing journals is on the left. The base map of 10,253 cited journals is on the right. The colors depict clusters identified by the Blondel clustering algorithm. (*See insert for color representation of the figure.*)

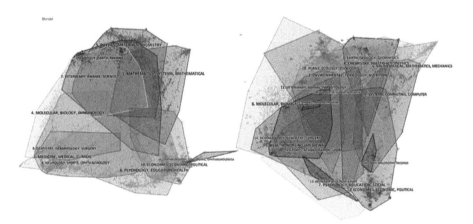

FIGURE 7.7   The boundary of each cluster is shown to depict how its members are distributed. Clusters in both base maps overlap substantially. (*See insert for color representation of the figure.*)

from the source journal to the target journal of the citation. The curve style is designed to improve the clarity of the visualization of a large number of citation links.

Clusters with less than five members are not shown labels. The label of a cluster is represented by terms selected from the titles of journals in the cluster. The label of a cluster is placed at the cluster centroid. As shown in Figure 7.7, the boundaries of

Blondel clusters in both base maps have considerable overlaps between multiple clusters. It is also clear that journals are not evenly distributed. Since cluster memberships are exclusive, reducing the amount of overlapping would be a preferable move if layout algorithms can effectively separate nodes in different clusters (Dwyer et al., 2008). However, a study of this issue is beyond the scope of this study. It should be noted that in this study we make no assumptions concerning the presence or absence of overlapping clusters.

7.3   EXAMPLES OF DUAL-MAP OVERLAYS

We demonstrate the use of dual-map overlays with examples of different types in terms of how a portfolio of publications is constructed. The first type is a *single-source overlay*, which represents portfolios that are generated with a single seed article, that is, a portfolio of this type consists of all the articles that cite the seed article. We will include one example of a single-source overlay on the topic of autism and vaccines. The second type is an *organizational overlay*, which represents a portfolio of an organization, including a department in a university, a corporate research lab, or a national laboratory. Examples of organizational overlays include publication portfolios of three iSchools in the United States and publication portfolios of three well-known corporations. The third type is *subject matter overlays*, which are defined by the relevance to an underlying subject matter. Examples of subject matters include regenerative medicine, mass extinctions, visual analytics, and articles that cited the *Journal of the American Society for Information Science and Technology* (*JASIST*).

These examples are representative of common types of scenarios in the context of portfolio analysis, and we will demonstrate prominent patterns revealed by the new method. For instance, the single-source overlay example is essentially originated from a single publication. The three corporations in the organizational overlay examples are widely known. Examples of subject matter overlays are topics we either have previously studied or are familiar with. The diversity of these examples illustrates the scope and flexibility of the new method.

7.3.1   Portfolios of a Single Source

In 1998, a paper by Wakefield et al. appeared in *Lancet* (Wakefield et al., 1998). It suggested a possible link between a combination of vaccines against measles, mumps, rubella, and autism. The study had drawn a considerable amount of attention from researchers and the general public. As a result, many parents in England decided to skip these vaccines for their children. In 2004, *Lancet* partially retracted the Wakefield paper. In 2010, the journal finally retracted the paper altogether. The Wakefield paper had been controversial for years prior to its retraction. The *Lancet*'s retraction notice in February 2010 noted that several elements of the 1998 paper were incorrect, contrary to the findings of an earlier investigation, and that the paper made false claims of an "approval" of the local ethics committee.

According to the Web of Science, the Wakefield paper is the most cited article that has been retracted. Its impact had evidently reached over 740 articles that cited the controversial paper. Some of the one-citation-away citing articles became highly cited themselves later on. For example, among the articles that cited Wakefield et al.'s study, one was cited by 384 articles and the other by 360 citations. Articles that cited Wakefield et al.'s article further amplified the influence of the later retracted study to an even broader context—the 740 one-citation-away articles were cited by more than 6600 articles in the Web of Science. These two-citation-away articles cited an even larger body of literature of over 12,000 references. The original paper's citation count peaked in 2002. A detailed analysis of citation contexts associated with retracted articles, including Wakefield article, can be found in our study of retracted scientific articles (Chen, Hu, Milbank, & Schultz, 2013).

A single-source overlay represents citation patterns of articles concerning a seed article. All the articles that cite the same seed articles are used to form the overlay. A seed article can be a groundbreaking article that represents a scientific breakthrough or a transformative discovery. A seed article could be a controversial or even retracted article of interest.

We explain one example of single-source overlays in detail here. The seed article in the example, Wakefield et al. (1998), is a highly cited retracted article, which has profound implications on public health, especially on vaccine uptakes from children.

Autism and Vaccines

We use the Wakefield paper as an example to illustrate various patterns that can be discerned from a dual-map overlay (Wakefield et al., 1998). The source of the overlay is a set of 405 articles that cited the Wakefield paper (this is a subset because of the limitation of our Web of Science subscription). Figure 7.8 shows the overlay with annotations to key points of interest. The bar charts near the top of the figure depict stepwise drifts in trajectories aggregated from the citing behavior of the 405 articles. The first bar on the left of the bar chart represents the amount of shift in 1999 with reference to the weight center of the disciplines involved in 1998. The chart shows that the distance of the shifts increased substantially between 2005–2006 and 2011–2012. Given that the Wakefield paper was partially retracted in 2004 and fully retracted in 2010, is it reasonable to hypothesize that a significant change in relevant scientific disciplines may attract new publications from new perspectives? New perspectives would result in publications in different journals.

Citations made by these source articles are shown as the spline waves, which are primarily in yellow, green, and cyan. Each spline curve starts from a citing journal in the base map on the left and points to a cited journal in the base map on the right. Labels in the vicinity of the launching areas indicate corresponding disciplines in which citing articles were published. Each label is centered at the cluster centroid of the corresponding journals. In this example, relevant disciplines include medicine, clinical, biology, immunology, psychology, education, and health on the citing side of the dual map. The majority of the citations were directed to disciplinary areas such as health, nursing, and medicine in the cited base map. Cocitation links that connect different disciplines can be displayed as dashed lines.

FIGURE 7.8 Citation patterns in an overlay of 405 articles that cited the Wakefield paper. (*See insert for color representation of the figure.*)

The lower half of the figure shows the trajectory of citing patterns on the left and the trajectory of cited patterns on the right. Properties of a citing trajectory can tell us about the dynamics of publications concerning the Wakefield paper at a disciplinary level. For example, if the citing trajectory shows a shift from one region to another on the base map of citing journals, we would know that there was a change of the primary disciplines in terms of relevant articles were published in a different set of journals. In this case, the citing trajectory is drifting toward the right-hand side of the citing base map. Based on the citation links shown in the upper half of the overlay map, the starting position of the citing trajectory is predominated by publications in the discipline of medicine and clinical medicine, whereas the ending position of the trajectory appears to be influenced by activities in areas near to the disciplines of psychology, education, and health.

The most active citing journal after 2004 is *Social Science Medicine*. For example, a paper published in 2005 in the journal reported an ethnographic study of the choice of vaccine (Poltorak, Leach, Fairhead, & Cassell, 2005): "In the context of the high-profile controversy that has unfolded in the UK around the measles, mumps and rubella (MMR) vaccine and its possible adverse effects, this paper explores how parents in Brighton, southern England, are thinking about MMR for their own children."

7.3.2   Portfolios of Organizations

The construction of an organizational overlay is based on a search in the address field in the Web of Science. For example, the portfolio of the College of Information Science and Technology[4] at Drexel University can be constructed by searching for bibliographic records that have the name of the college in the Address field. A portfolio of an individual researcher can be obtained by adding the author's name to the search.

Three iSchools
Publications with author affiliations involving one of the three iSchools in the United States are used as the source of three overlays, one for each iSchool. The window of analysis starts from 2003 and finishes at the end of 2012.

Three threads of citations stand out in Figure 7.9. The blue thread connects the cluster of mathematics and systems in the citing base map to the cluster of systems and computing in the cited base map. Representative journals of this thread include *Data and Knowledge Engineering*, *IEEE Computer Graphics and Applications*, and *IEEE Computer*. The two threads following the red lines are also prominent. The upper thread of the two essentially connects the library and information science in the citing journal base map to computing and information systems in the cited journal base map. Representative citing journals include *JASIST*, *Information Processing and Management*, and *Journal of Informetrics*. The lower thread of the two represents citations from journals such as *Journal of Computer-Mediated Communication*,

---

[4] The new name of the college is the College of Computing and Informatics after September 2013.

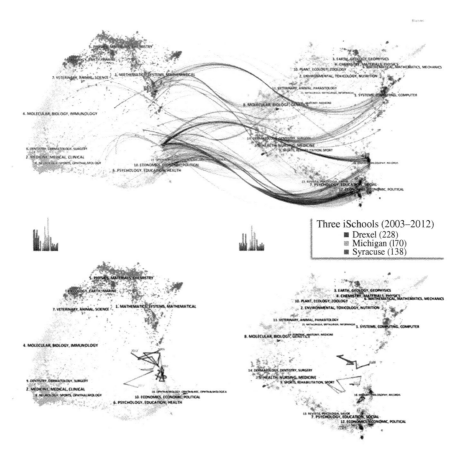

FIGURE 7.9 Overlays of three iSchools show major threads of citations that may characterize the publication portfolios of these institutions. The lower half of the figure shows the citing and cited trajectories in each of the base maps. (*See insert for color representation of the figure.*)

Computer Human Behavior, and *Government Information Quarterly* to the corresponding cluster in the cited base map. These disciplinary patterns provide useful insights into the nature of iSchools in terms of the core values perceived by the iSchools, namely, information, users, and technology.

The citing trajectories of the three iSchools have different patterns. The red trajectory had been concentrating on a small set of journals, whereas the blue and green trajectories had made long jumps across a larger group of journals. Their cited trajectories demonstrated similar patterns. In a portfolio analysis, one can investigate further to uncover the reasons that led to the differences at the global level and take actions accordingly. For example, if a trajectory with long jumps turns out to reflect the diverse expertise of new faculty members, then one may take this into account when considering hiring new faculty members. Multiple concentrations on the citing map may trace to the same source on the cited map. If that is the case, then authors of the seemingly diverse concentrations may share more common grounds than they probably realized.

The examples we present here are intended to illustrate the new method of portfolio analysis rather than present the results of a portfolio analysis because the data was collected from the Web of Science only. For an actual portfolio analysis, it is advisable to construct a comprehensive portfolio from multiple sources.

Three Corporations

In the following example, portfolios of publications by authors from three corporations, Google, Microsoft, and IBM, were constructed from bibliographic records in the Web of Science. Publications for each corporation were identified based on the address field of these bibliographic records. In order to concentrate on original research outputs from these corporations, we retained publications of type Article only for the study. Other types such as Review or Note were excluded.

Google's portfolio consisted of 620 source articles, which appeared in 550 unique journals and cited 8724 journals. Microsoft's portfolio had 1,968 papers from 1,050 journals and cited 21,193 journals. IBM's portfolio is the largest of the three, containing 3,965 articles from 1,593 journals and cited 27,617 journals (see Table 7.2).

Each corporation was assigned a distinct color: Google (blue), Microsoft (red), and IBM (green). These colors were used in their trajectories, bar charts, citation arcs, and cocitation links. As shown in Figure 7.10, the trajectories of all three organizations were located in the upper right region. In particular, IBM's trajectory (green) appeared slightly higher up in both base maps, relatively closer to disciplines such as physics and mathematics. Microsoft's and Google's trajectories appeared slightly lower in the map, relatively closer to psychology and other humanity-related disciplines.

The middle row in Figure 7.10 shows the trajectories of the three corporations. Citing trajectories are shown in the image on the left. Cited trajectories are on the right. Overall, the citing trajectories of Google and Microsoft appeared in areas near to each other, whereas the trajectory of IBM was farther away from them. At the global level, these patterns suggest that Google and Microsoft are more similar than IBM in terms of where they publish. Their trajectories in the cited map were separated apart evenly, indicating that they built on distinct areas of prior knowledge for their work.

The bar charts shown at the bottom of Figure 7.10 represent the length of each move in their trajectories. A short bar indicates a near-range move. A tall bar indicates a long-range move. All the three corporations made near-range moves in their citing trajectories, suggesting that they published in a relatively stable set of journals over

Table 7.2   Portfolios of three organizations' publications during 2008 and 2012

Organization	Color	Articles	Citing journals	Cited journals
Google	Blue	620	550	8,724
Microsoft	Red	1968	1050	21,193
IBM	Green	3965	1593	27,617

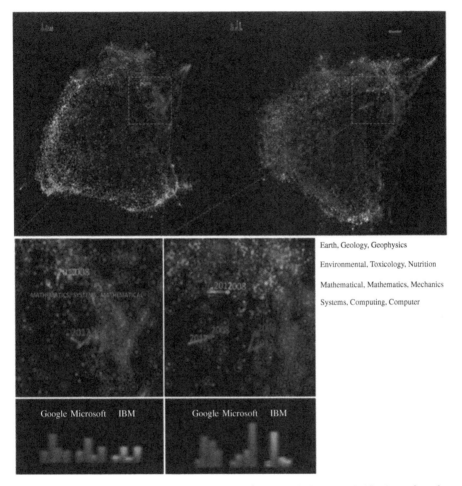

Earth, Geology, Geophysics

Environmental, Toxicology, Nutrition

Mathematical, Mathematics, Mechanics

Systems, Computing, Computer

FIGURE 7.10   Trajectories of Google (blue), Microsoft (red), and IBM (green). (*See insert for color representation of the figure.*)

the years. In contrast, their cited trajectories revealed longer jumps. For example, Microsoft made a substantial shift in the most recent year and IBM made a large shift in the second year of the window, which means they had changed substantially what they chose to cite.

Figure 7.11 shows the citation overlays of the three organizations. On top of the figure, in the base map of citing journals on the left, citations made by Google mostly originated from the area labeled as mathematics and systems (not shown in the figure, but accessible interactively). The majority of the citation arcs led to the corresponding area of the same discipline in the base map of cited journals on the right. The overlay in red below Google's overlay is from Microsoft. In addition to the same citation passageway as Google's citations, Microsoft's citation arcs followed a wider range of citation passageways. For example, Microsoft's citations reached several areas

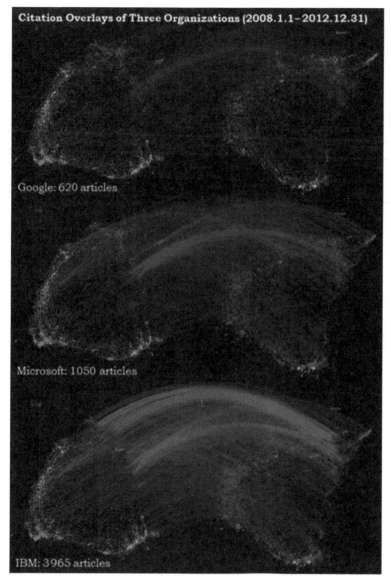

FIGURE 7.11   Citation overlays of three corporations. (*See insert for color representation of the figure.*)

located in the upper left part of the citing map, whereas these areas were not active in Google's portfolio. IBM's portfolio shows an even broader scope. The prominent trail of green arcs on the top of IBM's overlay chart highlighted some of the IBM's major competencies with more hardware-related areas. In contrast, this passageway was not prominent in the portfolio overlays of Google and Microsoft, which were more active in software-oriented areas.

With portfolio overlays and aggregated trajectories of organizations, one can quickly glean insightful patterns of these organizations. Furthermore, these patterns draw our attention to a subset of publications in a portfolio. We can then pursue more detailed information about publications associated with a particular pattern. Overlay and trajectory patterns at a macroscopic level provide a useful gateway to the study of the dynamics at both macroscopic and microscopic levels.

7.3.3   Portfolios of Subject Matters

A subject matter portfolio consists of publications relevant to the subject matter. Such portfolios can be constructed by a topic search in the Web of Science. A portfolio of research in regenerative medicine, for example, can be obtained by searching for bibliographic records relevant to "regenerative medicine" in the Web of Science.

Regenerative Medicine
Regenerative medicine is a rapidly growing area of research. It has many clinical implications and potentials. In a study published in 2012, we found that the topic of induced pluripotent stem cells (iPSCs) plays a leading role in regenerative medicine research (Chen, Hu, Liu, & Tseng, 2012). iPSCs research was awarded the 2012 Nobel Prize in Medicine. Figure 7.12 shows the trajectories of regenerative medicine. We updated the dataset with a new topic search for "regenerative medicine" between 2005 and 2012 in the Web of Science. A total of 3559 records found in this time frame were used to generate the overlay.

The bar charts of the trajectories, shown on the top of the figure, indicate that the trajectories are stable. The citing trajectory on the left closely tracked the disciplines along the disciplinary region labeled by terms such as molecular, biology,

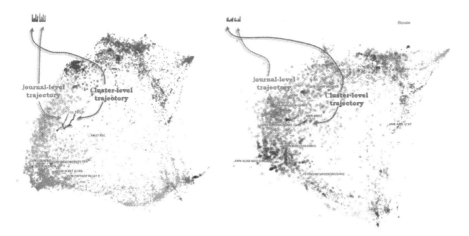

FIGURE 7.12   Trajectories of regenerative medicine research (2005–2012). The citing trajectory remains to be in the disciplinary area labeled as molecular, biology, and immunology throughout the entire course. (*See insert for color representation of the figure.*)

Table 7.3   Journals involved in the regenerative medicine dataset (2005–2012)

Year	2005	2006	2007	2008	2009	2010	2011	2012
Citing	70	81	144	192	222	265	325	319
Cited	1712	1912	2029	2321	2304	2379	2571	2618

and immunology. Throughout the entire course, the citing trajectory remains in the same discipline.

Table 7.3 lists the number of citing and cited journals per year in the regenerative medicine dataset. The few journals at both ends are likely to contribute to the instability of the trajectory.

Mass Extinctions (1975–2010)

Positions of trajectories in previous examples are calculated at the journal level. In this example, we calculate positions of trajectories at the disciplinary level. At a particular year, the positions of journals are matched to the cluster centroids of their corresponding clusters. Trajectories at the discipline level are expected to be more stable than trajectories at the journal level because many journal-to-journal movements within the same disciplinary cluster would be absorbed to a stable centroid of the same cluster.

Figure 7.13 shows the citing trajectory of mass extinctions research at the discipline level. The trajectory has a core discipline almost right at the center of the area labeled as ecology, earth, and marine. The trajectory spent most of the time in this area, except a long-range movement along the shape of a long triangle between 1977 and 1980. The longest distance it has moved was from the discipline labeled as physics, materials, and chemistry to the center of the region labeled as molecular, biology, and immunology. (See the schematic sketch on the top of Figure 7.13.) The long-distant move returned to the core of the trajectory next year. What specific papers caused the long-distant move? What kept the trajectory to such a compact core discipline for so many years? Studies of the structure and dynamics of specialties at a lower level of granularity are more appropriate to address this type of questions. For example, in a previous study of mass extinctions, we identified turning points in mass extinctions research (Chen, Cribbin, Macredie, & Morar, 2002). The overlay example here demonstrates how it may be integrated with the study of the dynamics of a specialty.

Visual Analytics (2006–2012)

The third example of a subject matter overlay is based on publications on visual analytics between 2006 and 2012. Figure 7.14 shows that the majority of visual analytics publications originated in the discipline of mathematics and computer science (threads in red originated from the red cluster in the citing base map on the left). The way in which visual analytics connects various disciplines was highlighted by cocitation links between disciplines—dashed lines connecting the centroids of clusters.

Cocitation links between clusters of cited journals show that visual analytics is primarily drawn upon the work in disciplines such as (1) computing and systems, (2) psychology and sociology, (3) economics and politics, and (4) plant, ecology, and zoology.

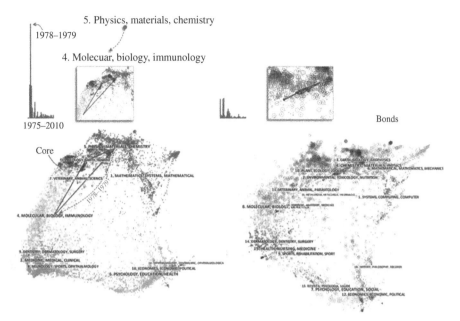

FIGURE 7.13 Trajectories of research in mass extinctions (1975–2010) at the discipline level. The core discipline of the research is identified as the Blondel cluster 3 on ecology, earth, and marine. The longest single-year shift occurred between 1978 and 1979 as the disciplinary center of the journals shifted from the Blondel cluster 5 on physics, materials, and chemistry to the Blondel cluster 4 on molecular, biology, and immunology. (*See insert for color representation of the figure.*)

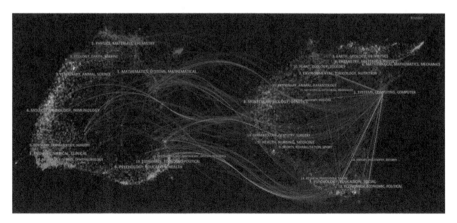

FIGURE 7.14 An overlay of publications in visual analytics (2006–2012). Wavelike curves depict citation links. They are colored by their source clusters. Dashed lines depict cocitation links across disciplinary boundaries. (*See insert for color representation of the figure.*)

Articles Citing JASIST Publications (2002–2011)

Articles that cite the *JASIST* between 2002 and 2011 were retrieved from the Web of Science (Table 7.4). An overlay was generated to reveal the impact of the journal. Figure 7.15 shows the same overlay in two different styles. The style used in the

Table 7.4   The number of journals involved in articles citing *JASIST*

	2002	2003	2004	2005	2006	2007	2008	2009	2010	2011
Citing journals	8	40	61	100	131	179	250	336	374	439
Cited journals	58	289	655	999	1480	1763	2211	3176	3230	3711

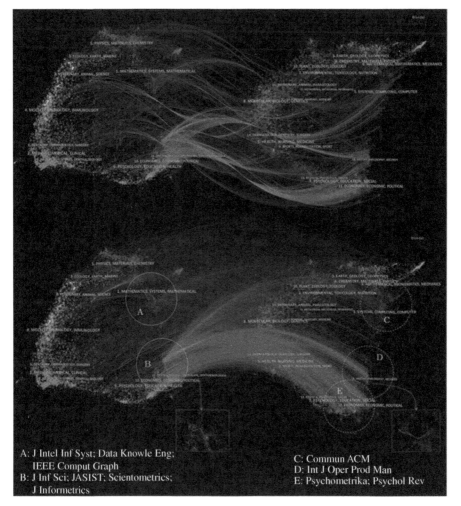

A: J Intel Inf Syst; Data Knowle Eng;
  IEEE Comput Graph
B: J Inf Sci; JASIST; Scientometrics;
  J Informetrics

C: Commun ACM
D: Int J Oper Prod Man
E: Psychometrika; Psychol Rev

FIGURE 7.15   An overlay of articles citing *JASIST* (2002–2011). (*See insert for color representation of the figure.*)

upper overlay depicts citation links in spline curves, whereas the style used in the lower overlay shows citation links in arcs. Both styles of citation links are colored by the corresponding source clusters of journals.

The overlay map shows that two areas on the citing map are particularly active, namely, area A (computer science) and area B (information science). The most frequently seen journals in area A include *Journal of Intelligent Information Systems*, *Data and Knowledge Engineering*, and *IEEE Computer Graphics and Applications*. The top journals in area B include *Journal of Informetrics*, *JASIST*, and *Scientometrics*. The citation arcs reveal three areas C, D, and E. The patterns revealed by citation arcs connecting disciplinary areas in the two base maps are straightforward to interpret once the user becomes familiar with the "geography" of the base maps.

7.3.4 Patterns in Trajectories

Trajectories visualized with the dual-map overlays enable the study of the dynamics of a complex system characterized by a portfolio as a whole. Some of the patterns may provide useful signs to guide more in-depth pursuits at lower levels of granularity. A trajectory is a sequence of disciplines or journals that are most representative of the collective publishing behavior of scientists. Trajectories can be formed at the level of individual journals and at the level of clusters of journals to represent disciplines. We will focus on the disciplinary trajectories in the rest of the section, although the metrics and interpretations can be translated to the journal-level trajectories.

The set of unique disciplines in a trajectory provides useful information to differentiate different trajectories. For example, two trajectories may involve the same set of disciplines. Some trajectories may essentially deal with one discipline, whereas other trajectories may deal with a large number of disciplines. Given the same set of disciplines, trajectories may differ in terms of the order these disciplines appear and how often each individual discipline appears.

We define the distance between two disciplines in terms of the distance between the centroids of the two disciplines. The centroid of a discipline is defined as the weight center of the journals that belong to the same Blondel cluster. The proximity of two disciplines on the base maps reflects the strength of the interdisciplinary connection between them because the base maps are built on citing and cited patterns of journals. Using a combination of the disciplines involved and the proximity of these disciplines, we can derive a number of patterns.

Core Disciplines—If a trajectory either repeatedly enters or persistently remains in the same discipline, it may represent the collective behavior of scientists who mostly publish in the same disciplines.

Multiple Disciplines with Variations of Proximity—A trajectory may involve multiple disciplines. These multiple disciplines may be traversed in a variety of order. The proximity of adjacent disciplines may change slowly or rapidly. For example, a rapid change of proximity indicates a long-range move or a shift. A small step in a trajectory represents a drift or a short-range move.

Closed and Open—A closed trajectory may drift and/or shift back and forth multiple times between disciplines in a small group, whereas in an open trajectory, the number of disciplines involved is constantly growing. Since the total number of Blondel clusters in our study is fixed, an open trajectory is relatively speaking. If a scientific breakthrough is made and it has a broad impact on other disciplines, we would expect that the number of disciplines involved will increase rapidly in the following years. On the other hand, the trajectory of a research program addressing a highly unique problem may be limited to a small number of disciplines and show a closed trajectory.

We illustrate some of these patterns with disciplinary trajectories of two productive and high-impact scholars, Edward Witten, a physicist, and Ben Shneiderman, a computer scientist. Both of them have published hundreds of articles in their fields. According to the Web of Science, Witten has 228 publications indexed by the Science Citation Index (SCI) with an h-index of 115. The 228 publications have been cited by 37,523 records without self-citations. His average citation per item is 279.50. Shneiderman has 156 publications in the Web of Science with an h-index of 30. The 156 publications have been cited by 2489 records with an average of 18.67 citations per item. Note that these statistics are by no means comprehensive and they are likely to be biased by the source of data. For example, Google Citation Profile lists of an h-index of 89 and 46,324 citations for Shneiderman, whereas Witten does not have set up his Google Citation Profile.

We retrieved bibliographic records of the type article from the Web of Science for each of them. The article type corresponds to original research publications. 206 articles and 132 articles were retrieved for Witten and Shneiderman, respectively. Trajectories were generated at both discipline and journal levels. Figure 7.16 illustrates the trajectories of the overlays generated from the two sets of publications. Witten's publications projected a compact and stable citing trajectory in areas

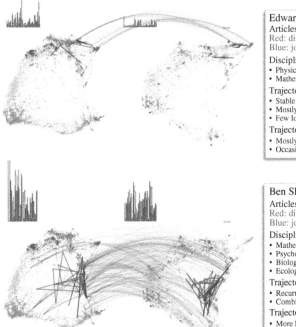

Edward Witten
Articles: 206 (1980–2011)
Red: discipline level
Blue: journal level
Disciplines:
• Physics (29 times)
• Mathematics (3 times)
Trajectory of Citing Journals
• Stable and compact core
• Mostly short-range shifts
• Few long-range shifts
Trajectory of Cited Journals
• Mostly near-range shifts
• Occasionally short-range shifts

Ben Shneiderman
Articles: 132 (1980–2012)
Red: discipline level
Blue: journal level
Disciplines:
• Mathematics/Computing (17 times)
• Psychology (8 times)
• Biology (2 times)
• Ecology (1 times)
Trajectory of Citing Journals
• Recurring multiple cores
• Combined short and long-range shifts
Trajectory of Cited Journals
• More long-range shifts and short-range ones
• No liner trend

FIGURE 7.16   Characteristics of trajectories of Witten's publications and Shneiderman's publications. Citing trajectories of overlays are shown on the left. Cited trajectories are shown on the right. (*See insert for color representation of the figure.*)

corresponding to physics. The moves of Witten's trajectory are mostly short-range moves. The journal-level trajectory shows some brief episodes of publications in mathematics. No other disciplines were involved. In contrast, the citing trajectory of Shneiderman shows that he routinely published in two disciplines (mathematics/computing and psychology/education/health) and occasionally published in another two disciplines (biology and ecology).

Table 7.5 lists the nearest disciplines of the citing trajectories of both scholars over time. The names of disciplines are represented by terms chosen from titles of journals in the corresponding Blondel cluster of citing journals. Witten's publications

Table 7.5 Citing trajectories at the discipline level

	Disciplines of publications (Blondel cluster labels)	
Year	Edward Witten's publications	Ben Shneiderman's publications
1980	Physics, Materials, Chemistry	Psychology, Education, Health
1981	Physics, Materials, Chemistry	Mathematics, Systems, Mathematical
1982	Physics, Materials, Chemistry	Mathematics, Systems, Mathematical
1983	Physics, Materials, Chemistry	Mathematics, Systems, Mathematical
1984	Physics, Materials, Chemistry	
1985	Physics, Materials, Chemistry	Psychology, Education, Health
1986	Physics, Materials, Chemistry	Mathematics, Systems, Mathematical
1987	Physics, Materials, Chemistry	Psychology, Education, Health
1988	Physics, Materials, Chemistry	Mathematics, Systems, Mathematical
1989	Physics, Materials, Chemistry	
1990	Physics, Materials, Chemistry	
1991	Mathematics, Systems, Mathematical	Mathematics, Systems, Mathematical
1992	Physics, Materials, Chemistry	Mathematics, Systems, Mathematical
1993	Physics, Materials, Chemistry	Mathematics, Systems, Mathematical
1994	Physics, Materials, Chemistry	Mathematics, Systems, Mathematical
1995	Physics, Materials, Chemistry	Mathematics, Systems, Mathematical
1996	Physics, Materials, Chemistry	
1997	Physics, Materials, Chemistry	Mathematics, Systems, Mathematical
1998	Physics, Materials, Chemistry	Psychology, Education, Health
1999	Physics, Materials, Chemistry	Mathematics, Systems, Mathematical
2000	Physics, Materials, Chemistry	Mathematics, Systems, Mathematical
2001	Physics, Materials, Chemistry	Psychology, Education, Health
2002	Physics, Materials, Chemistry	Mathematics, Systems, Mathematical
2003	Physics, Materials, Chemistry	Molecular, Biology, Immunology
2004	Physics, Materials, Chemistry	Psychology, Education, Health
2005	Physics, Materials, Chemistry	Ecology, Earth, Marine
2006	Physics, Materials, Chemistry	Mathematics, Systems, Mathematical
2007	Mathematics, Systems, Mathematical	Mathematics, Systems, Mathematical
2008	Mathematics, Systems, Mathematical	Psychology, Education, Health
2009	Physics, Materials, Chemistry	Mathematics, Systems, Mathematical
2010	Physics, Materials, Chemistry	
2011	Physics, Materials, Chemistry	Molecular, Biology, Immunology
2012		Psychology, Education, Health

are remarkably persistent, containing an almost noninterrupted stream of the same discipline, physics. In contrast, Shneiderman's publication trajectory has shown an oscillation between mathematics and computing, and psychology, education, and health. Blank rows in the table indicate that the Web of Science has no records of Shneiderman's publication indexed during that year, which doesn't indicate whether or not he published outside the coverage of the Web of Science. Shneiderman is widely known for his work on human–computer interaction; thus, we expect the recurring psychology/education/health reflects his work in that aspect. He is also well known for his work in information visualization and other computing areas, which would correspond to the recurring mathematics, systems, and mathematical in his trajectory. More interestingly, the trajectory appears to merge two streams together such that one episode of one discipline is followed by an episode of another discipline. The switch between the two disciplines almost appears to be periodic. For example, the trajectory has repeated the shifts from psychology, education, and health to mathematics and systems for six times since 1980.

The contrast between the two illustrative examples shows that one can discern useful information from trajectories of portfolios of individual scientists at disciplinary and journal levels. Examining further details at lower levels of granularity is feasible, especially in connection with more traditional scientometric studies. The same type of analysis of trajectories can be used to study portfolios of a wide range of units of analysis at organizational and international levels.

7.4   SUMMARY

We have demonstrated the potential of simultaneously displaying two global maps of science at the discipline level. The dual-map design enables an explicit, intuitive, and easy-to-interpret representation of citations made by a wide variety of portfolios of publications. The dual-map space provides a flexible and extensible framework to support a new set of visual analytic tasks that are essential for portfolio analysis, gap analysis, and competitive intelligence. The notion of an aggregated trajectory of a portfolio provides an additional new gateway from the study of macroscopic patterns to the dynamics at microscopic levels.

Several issues need to be addressed and have room for improvement in the future. One issue that we have not addressed in the development of the dual-map overlay design is the stability of global science maps, and how their stability would influence the validity of patterns revealed. Pragmatically, how often do we need to update the underlying base maps in order to maintain the reliability of patterns of an overlay? Although the issue is concerned with science mapping in general, the role increasingly played by thematic overlays in portfolio analysis highlights the need to investigate the issue in particular. Another issue is related to the layout of the base maps. Our visualization has revealed a substantial amount of overlaps among Blondel clusters in both citing and cited base maps. Future research may investigate feasible trade-offs between the layout of base maps and their role as a gateway to integrate analytical tasks at various levels of granularity.

Our examples have demonstrated the flexibility of global maps of science at the level of journals and clusters of journals. A related issue is to what extent the new method introduced here can be applied to other types of global maps of science, such as a global map of science constructed at higher or lower levels of granularity than journals. In particular, a topic map of science derived from promising techniques such as topic modeling may apply. Leydesdorff and his colleagues have extended the base map construction process from scholarly publications to patents (Leydesdorff, Kushnir, & Rafols, 2014). The method described here is applicable to a dual-map overlay of patent portfolio analysis and even to a hybrid publication and patent portfolio analysis. We are actively pursuing an extension of the dual-map method to patent portfolio analysis. Our experience with the dual-map overlays also suggests that it may be worth considering multimap overlays for a comprehensive portfolio analysis that may involve multiple types of entities and relations, such as publications, patents, and grants.

The dual-map design enables analysts to perform several new and intuitive types of visual analytic tasks for portfolio analysis, including comparing dynamic patterns and trends of multiple portfolios at multiple levels of granularity from individual citation arcs, to dynamic patterns of trajectories aggregated over an entire portfolio. The dual-map design provides a new conceptual framework in which one can derive a variety of new metrics and algorithms. For example, we have introduced the concept of structural variation and its implications on detecting and predicting potentially transformative contributions of scientific publications in the framework of a complex adaptive system (Chen, 2012). The dual-map design provides an opportunity to study the predictive effects of structural variation from an alternative perspective. We will pursue this opportunity in our subsequent studies.

We have introduced a new method for portfolio analysis based on a dual-map design. The potential of the method is demonstrated through a series of examples of a variety of portfolios of publications, ranging from individual scientists, organizations, and subject matter-focused fields of research. We have shown that multiple overlays on the dual-map visualization can facilitate the analysis of portfolios in terms of identifying the areas of competencies and patterns of movements with reference to multiple disciplines. The dual-map overlays provide an intuitive gateway to integrate the study of scientific disciplines at a macroscopic level and the study of more specific specialties at a lower level of granularity. We can expect that the new method may lead to fruitful routes of further research and enrich the available methodologies for portfolio analysis, gap analysis, and competitive intelligence.

7.5 CONCLUSION

We begin the book with questions on what attracts our attention and what makes something interesting. A gap between what we know and what we don't know has emerged as one of the major driving forces behind answers to these questions. The magnitude of these gaps matters. If the gap is too small, then we are likely to use the perspective that we are currently using to answer the question. If the gap is too large, then we may not even consider making any effort to bridge the gap. With a gap of an "interesting" length, we are attracted to learn and potentially find a new perspective to see the world.

In Chapter 2, we devoted to the role of mental models in guiding our thinking and decision making and discussed the risks of having a wrong mental model. Ironically, it often takes little effort to build up a mental model, but it may take much more effort and much stronger evidence to make us realize that we probably ought to update or discard our current mental model. We take many things for granted. Sometimes, it may be one thing too many.

In Chapter 3, we consolidated the mental model theme further through a series of examples that underlined the subjectivity of evidence. The same information may serve as evidence for different purposes and may support different arguments. We characterized the ambiguity with a metaphor of the tips of quite different icebergs. The examples of making arguments to prove or disprove that a proposition is beyond the reasonable doubt highlighted the challenges we face when we deal with the complexity of various plausible scenarios and our beliefs.

In Chapter 4, we summarized the CiteSpace system that we have developed to reduce the level of complexity in understanding the structure and dynamics of scientific knowledge. CiteSpace supports a generic process of systematically analyzing the structure of a complex adaptive system.

In Chapter 5, we reviewed the origin of the fitness landscape paradigm in evolutionary genetics and its subsequent variations and extensions beyond biology. The evolutionary thinking behind fitness landscapes is one of the few recurring themes throughout the book. Evolution offers an inspirational framework to rethink the nature of explorations, optimizations, recombinant search, and many profound phenomena.

In Chapter 6, we turned our attention to the exploratory, optimal, and recombinant search mechanisms manifested in inventing new ideas in high-impact radical patterns. We included several detailed examples to highlight the wisdoms of how to be creative. The implications of evolution theories are evident in examples that have gone beyond the realm of the genotype and phenotype. The notion of gaps between where we are and where we want to be is illustrated in generalized fitness landscapes.

In Chapter 7, we introduced a dual-map overlay technique for a new type of portfolio analysis. The concepts of landscapes, trajectories, and movements are integrated to reduce the complexity of detecting early signs of critical transitions in a complex adaptive system to trajectories that one can monitor and track as closely as possible.

There are a few grand challenges ahead. If we can resolve these challenges, then we will be able to make tremendous progress in improving our ability to handle the complexity of reality and a variety of issues concerning a complex adaptive system.

The first grand challenge is reducing the complexity of the information we need to deal with. How can we cut to the chase in assimilating a large amount of complex, uncertain, incomplete, and ambiguous information? Is it possible to reduce the complexity of how we represent and communicate scientific knowledge today? Fitness landscapes suggest a potentially fruitful direction to pursue because fitness landscapes encourage us to hide various details but direct our focus to where we are and what we can do to achieve our goals.

The second grand challenge is the increased preparedness for cognitive misperception. We are often unprepared for many things we take for granted. The worst-case scenario is that we are unprepared but we thought we were well prepared because we had the wrong mental model. What can we do to increase our consciousness that there may be alternative interpretations of what we see? How can we reduce the risk of missing subtle but critical warning signs?

The third grand challenge is to be able to detect early signs of critical transitions in a complex adaptive system. This is probably the most challenging area and also the one that we will benefit the most from if the challenge is resolved.

What these grand challenges have in common is the fitness of information. We have visited a number of principles, procedures, and concrete examples in this book. The theme of the fitness of information represents itself in many variations, notably including the genetic fitness of a population based on combinations of genes at the genotypic level, the fitness of a compound as a drug candidate as a function of its molecular structure, the fitness of a country in terms of its life expectancy as a function of its multidimensional array of economic indicators, and the fitness of an idea in terms of its predictive potential as an early warning sign for a critical transition in scientific knowledge. This diverse range of ways to characterize the fitness of information underlines the fundamental value of an evolutionary paradigm in dealing with the complexity of data science. The fitness of information leads naturally to the notion of a generalized fitness landscape and a tightly coupled relationship between the information we are analyzing and actions we may pursue to accomplish our long-term goals.

Figure 7.17 illustrates an ideal fitness landscape of scientific information. Many scientific inquiries can be conceptualized as a creative search on such a landscape. In some areas, scientists have consistent findings. In other areas, there may be con-tradictions, such as competing paradigms and conflicting mental models. Some areas may be well defined and populated, whereas other areas may be ambiguous, uncertain, or totally uncharted. There are peaks resulted from the Matthew Effect as well as valleys and no man's land where groundbreaking ideas may rise. The fitness

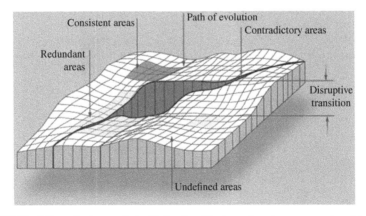

FIGURE 7.17 A fitness landscape of scientific inquires. (*See insert for color representation of the figure.*)

landscape provides an analytic framework for studying the behavior of a complex adaptive system at both microscopic and macroscopic levels. Ideally, the fitness landscape should provide an accurate representation of the history, the state of the art, and foresight of science. To paraphrase IARPA's RFI, what is the probability of having a fitness landscape of science of this kind on our desktop within five years?

In this book, we have explored and examined several important aspects of measuring the fitness of information in a diverse range of settings. The fitness value of information derived from complex, incomplete, ambiguous, and heterogeneous streams of data plays an essential role in maintaining our awareness of an evolving situation. The fitness of information also lets us focus on the basis of an evolutionary framework for gap analytics and portfolio analysis in a wide variety of domains.

The value of an evolutionary perspective is that it allows us to assess the fitness of information in the context of an optimization process, which can be a recombinant search, a boundary-spanning synthesis, a gap-bridging theory, a successive hill-climbing trajectory, or an exploration across no man's land.

As the tension increases between the size of the information haystack and the types of needles that may hold the key to critical transitions in a complex adaptive system, a focus on the fitness of information provides a good place to start.

BIBLIOGRAPHY

Blondel, V. D., Guillaume, J. L., Lambiotte, R., & Lefebvre, E. (2008). Fast unfolding of communities in large networks. *Journal of Statistical Mechanics: Theory and Experiment, 8*(10), 10008.

Chen, C. (2012). Predictive effects of structural variation on citation counts. *Journal of the American Society for Information Science and Technology, 63*(3), 431–449.

Chen, C. M. (2006). CiteSpace II: Detecting and visualizing emerging trends and transient patterns in scientific literature. *Journal of the American Society for Information Science and Technology, 57*(3), 359–377.

Chen, C., Cribbin, T., Macredie, R., & Morar, S. (2002). Visualizing and tracking the growth of competing paradigms: Two case studies. *Journal of the American Society for Information Science and Technology, 53*(8), 678–689.

Chen, C., Hu, Z., Liu, S., & Tseng, H. (2012). Emerging trends in regenerative medicine: A scientometric analysis in CiteSpace. *Expert Opinions on Biological Therapy, 12*(5), 593–608.

Chen, C., Hu, Z., Milbank, J., & Schultz, T. (2013). A visual analytic study of retracted articles in scientific literature. *Journal of the American Society for Information Science and Technology, 64*(2), 234–253.

Chen, C., Ibekwe-SanJuan, F., & Hou, J. (2010). The structure and dynamics of co-citation clusters: A multiple-perspective co-citation analysis. *Journal of the American Society for Information Science and Technology, 61*(7), 1386–1409.

Dwyer, T., Marriott, K., Schreiber, F., Stuckey, P. J., Woodward, M., & Wybrow, M. (2008). Exploration of networks using overview + detail with constraint-based cooperative layout. *IEEE Transactions on Visualization and Computer Graphics, 14*(6), 1293–1300.

Garfield, E., Pudovkin, A. I., & Istomin, V. S. (2003). Why do we need algorithmic historiography? *Journal of the American Society for Information Science and Technology*, *54*(5), 400–412.

Jin, B., & Rousseau, R. (2001). An introduction to the Barycenter method with an application to China's mean centre of publication. *Libri*, *51*(4), 225–233.

Leydesdorff, L., & Rafols, I. (2011). Local Emergence and Global Diffusion of Research Technologies: An exploration of patterns of network formation. *Journal of the American Society for Information Science and Technology*, *62*(5), 846–860.

Leydesdorff, L., Kushnir, D., & Rafols, I. (2014). Interactive Overlay Maps for US Patent (USPTO) data based on International Patent Classifications (IPC). *Scientometrics*, *98*(3), 1583–1599.

Leydesdorff, L., Rafols, I., & Chen, C. (2013). Interactive overlays of journals and the measurement of interdisciplinarity on the basis of aggregated journal-journal citations. *Journal of the American Society for Information Science and Technology*, *64*(12), 2573–2586.

Lucio-Arias, D., & Leydesdorff, L. (2008). Main-path analysis and path-dependent transitions in HistCite (TM)-based historiograms. *Journal of the American Society for Information Science and Technology*, *59*(12), 1948–1962.

Mackinlay, J. D., Rao, R., & Card, S. K. (1995). *An organic user interface for searching citation links*. Paper presented at the SIGCHI'95, May 7–11, 1995, Denver, CO.

Poltorak, M., Leach, M., Fairhead, J., & Cassell, J. (2005). 'MMR talk' and vaccination choices: An ethnographic study in Brighton. *Social Science Medicine*, *61*(3), 709–719.

Porter, A. L., & Rafols, I. (2009). Is science becoming more interdisciplinary? Measuring and mapping six research fields over time. *Scientometrics*, *81*(3), 719–745.

Rafols, I., Porter, A. L., & Leydesdorff, L. (2010). Science overlay maps: A new tool for research policy and library management. *Journal of the American Society for Information Science and Technology*, *61*(9), 1871–1887.

Wakefield, A. J., Murch, S. H., Anthony, A., Linnell, J., Casson, D. M., Malik, M., et al. (1998). Ileal-lymphoid-nodular hyperplasia, non-specific colitis, and pervasive developmental disorder in children (Retracted article. See vol 375, pg 445, 2010). *The Lancet*, *351*(9103), 637–641.

Waltman, L., & Van Eck, N. J. (2013). A smart local moving algorithm for large-scale modularity-based community detection. *European Physical Journal B*, *86*(11), 471.

Wright, S. (1932). The roles of mutation, inbreeding, crossbreeding, and selection in evolution. In: *Proceedings of the sixth international congress on genetics* (pp. 356–366), August 1932, Ithaca, NY.

Zhang, R., Zhang, Y., Zhang, Q., Xie, H., Qian, W., & Wei, F. (2013). Growth of half-meter long carbon nanotubes based on Schulz–Flory distribution. *ACS Nano*, *7*(7), 6156–6161.

Index

adaptation, 153–7, 167
adaptive landscape, 149–53, 155–6, 158, 161, 189–91, 227
adaptive walk, 190
ambiguity, 11, 105, 124, 177, 248
Anglo-American system of evidence, 77
Analysis of competing hypotheses (ACH), 42
arousal, 4
attention, 1–3, 7, 18, 37–8
attention space, 48
atypical combination, 185
atypical connection, 185, 187–8
author cocitation analysis (ACA), 123
automatic cluster labeling, 126
averageness hypothesis, 17

base map, 117–18, 160–161, 204, 219, 225, 228–31, 235, 247
Bayesian classifier, 90
Bayesian inference, 73, 78, 84–5
Bayesian network, 73–4, 87
Bayesian probability, 80, 83
Bayesian search, 73
beauty, 15–16
betweenness centrality, 124, 197
bibliographic coupling, 107, 128
bifurcation, 174
biological evolution, 51
black swan, 173–4
Blondel cluster, 227–8

Boston Marathon bombing, 7
boundary object, 32
boundary-spanning connection, 180
boundary-spanning mechanism, 55, 180, 188, 193
brokerage theory of discovery, 55
burst detection, 125, 222

carbon nanotube, 217–18, 229
centrality divergence, 195, 197
chess player, 10
chunking, 10
citation, 106
    analysis, 106
    indexing, 106, 127, 175
CiteSpace, 116–17, 119, 132, 140–141
closure principle, 6, 8
cluster labeling, 122
cluster linkage (CL), 195
clustering, 125
    Blondel, 228
    spectral, 122, 126, 195
cocitation analysis, 107
cocitation network, 99, 110, 122, 125, 194
cognitive map, 147
competing hypothesis, 41
complex adaptive system (CAS), 119, 173, 177–8, 192, 211
complex network analysis, 201
Composite Portraiture, 16

The Fitness of Information: Quantitative Assessments of Critical Evidence, First Edition. Chaomei Chen.
© 2014 John Wiley & Sons, Inc. Published 2014 by John Wiley & Sons, Inc.

255

WILEY SERIES IN PROBABILITY AND STATISTICS
ESTABLISHED BY WALTER A. SHEWHART AND SAMUEL S. WILKS

Editors: *David J. Balding, Noel A. C. Cressie, Garrett M. Fitzmaurice, Geof H. Givens, Harvey Goldstein, Geert Molenberghs, David W. Scott, Adrian F. M. Smith, Ruey S. Tsay, Sanford Weisberg*
Editors Emeriti: *J. Stuart Hunter, Iain M. Johnstone, Joseph B. Kadane, Jozef L. Teugels*

The *Wiley Series in Probability and Statistics* is well established and authoritative. It covers many topics of current research interest in both pure and applied statistics and probability theory. Written by leading statisticians and institutions, the titles span both state-of-the-art developments in the field and classical methods.

Reflecting the wide range of current research in statistics, the series encompasses applied, methodological and theoretical statistics, ranging from applications and new techniques made possible by advances in computerized practice to rigorous treatment of theoretical approaches.

This series provides essential and invaluable reading for all statisticians, whether in academia, industry, government, or research.

*Now available in a lower priced paperback edition in the Wiley Classics Library.
†Now available in a lower priced paperback edition in the Wiley–Interscience Paperback Series.

BARTOSZYNSKI and NIEWIADOMSKA-BUGAJ · Probability and Statistical Inference, *Second Edition*

BASILEVSKY · Statistical Factor Analysis and Related Methods: Theory and Applications

BATES and WATTS · Nonlinear Regression Analysis and Its Applications

BECHHOFER, SANTNER, and GOLDSMAN · Design and Analysis of Experiments for Statistical Selection, Screening, and Multiple Comparisons

BEH and LOMBARDO · Correspondence Analysis: Theory, Practice and New Strategies

BEIRLANT, GOEGEBEUR, SEGERS, TEUGELS, and DE WAAL · Statistics of Extremes: Theory and Applications

BELSLEY · Conditioning Diagnostics: Collinearity and Weak Data in Regression

† BELSLEY, KUH, and WELSCH · Regression Diagnostics: Identifying Influential Data and Sources of Collinearity

BENDAT and PIERSOL · Random Data: Analysis and Measurement Procedures, *Fourth Edition*

BERNARDO and SMITH · Bayesian Theory

BHAT and MILLER · Elements of Applied Stochastic Processes, *Third Edition*

BHATTACHARYA and WAYMIRE · Stochastic Processes with Applications

BIEMER, GROVES, LYBERG, MATHIOWETZ, and SUDMAN · Measurement Errors in Surveys

BILLINGSLEY · Convergence of Probability Measures, *Second Edition*

BILLINGSLEY · Probability and Measure, *Anniversary Edition*

BIRKES and DODGE · Alternative Methods of Regression

BISGAARD and KULAHCI · Time Series Analysis and Forecasting by Example

BISWAS, DATTA, FINE, and SEGAL · Statistical Advances in the Biomedical Sciences: Clinical Trials, Epidemiology, Survival Analysis, and Bioinformatics

BLISCHKE and MURTHY (editors) · Case Studies in Reliability and Maintenance

BLISCHKE and MURTHY · Reliability: Modeling, Prediction, and Optimization

BLOOMFIELD · Fourier Analysis of Time Series: An Introduction, *Second Edition*

BOLLEN · Structural Equations with Latent Variables

BOLLEN and CURRAN · Latent Curve Models: A Structural Equation Perspective

BONNINI, CORAIN, MAROZZI and SALMASO · Nonparametric Hypothesis Testing: Rank and Permutation Methods with Applications in R

BOROVKOV · Ergodicity and Stability of Stochastic Processes

BOSQ and BLANKE · Inference and Prediction in Large Dimensions

BOULEAU · Numerical Methods for Stochastic Processes

* BOX and TIAO · Bayesian Inference in Statistical Analysis

BOX · Improving Almost Anything, *Revised Edition*

* BOX and DRAPER · Evolutionary Operation: A Statistical Method for Process Improvement

BOX and DRAPER · Response Surfaces, Mixtures, and Ridge Analyses, *Second Edition*

BOX, HUNTER, and HUNTER · Statistics for Experimenters: Design, Innovation, and Discovery, *Second Editon*

BOX, JENKINS, and REINSEL · Time Series Analysis: Forcasting and Control, *Fourth Edition*

BOX, LUCEÑO, and PANIAGUA-QUIÑONES · Statistical Control by Monitoring and Adjustment, *Second Edition*

* BROWN and HOLLANDER · Statistics: A Biomedical Introduction

CAIROLI and DALANG · Sequential Stochastic Optimization

CASTILLO, HADI, BALAKRISHNAN, and SARABIA · Extreme Value and Related Models with Applications in Engineering and Science

CHAN · Time Series: Applications to Finance with R and S-Plus®, *Second Edition*

---

*Now available in a lower priced paperback edition in the Wiley Classics Library.
†Now available in a lower priced paperback edition in the Wiley–Interscience Paperback Series.

CHARALAMBIDES · Combinatorial Methods in Discrete Distributions
CHATTERJEE and HADI · Regression Analysis by Example, *Fourth Edition*
CHATTERJEE and HADI · Sensitivity Analysis in Linear Regression
CHEN · The Fitness of Information: Quantitative Assessments of Critical Evidence
CHERNICK · Bootstrap Methods: A Guide for Practitioners and Researchers,
  Second Edition
CHERNICK and FRIIS · Introductory Biostatistics for the Health Sciences
CHILÈS and DELFINER · Geostatistics: Modeling Spatial Uncertainty, *Second
  Edition*
CHIU, STOYAN, KENDALL and MECKE · Stochastic Geometry and Its
  Applications, *Third Edition*
CHOW and LIU · Design and Analysis of Clinical Trials: Concepts and
  Methodologies, *Third Edition*
CLARKE · Linear Models: The Theory and Application of Analysis of Variance
CLARKE and DISNEY · Probability and Random Processes: A First Course with
  Applications, *Second Edition*
* COCHRAN and COX · Experimental Designs, *Second Edition*
COLLINS and LANZA · Latent Class and Latent Transition Analysis: With
  Applications in the Social, Behavioral, and Health Sciences
CONGDON · Applied Bayesian Modelling, *Second Edition*
CONGDON · Bayesian Models for Categorical Data
CONGDON · Bayesian Statistical Modelling, *Second Edition*
CONOVER · Practical Nonparametric Statistics, *Third Edition*
COOK · Regression Graphics
COOK and WEISBERG · An Introduction to Regression Graphics
COOK and WEISBERG · Applied Regression Including Computing and Graphics
CORNELL · A Primer on Experiments with Mixtures
CORNELL · Experiments with Mixtures, Designs, Models, and the Analysis of
  Mixture Data, *Third Edition*
COX · A Handbook of Introductory Statistical Methods
CRESSIE · Statistics for Spatial Data, *Revised Edition*
CRESSIE and WIKLE · Statistics for Spatio-Temporal Data
CSÖRGŐ and HORVÁTH · Limit Theorems in Change Point Analysis
DAGPUNAR · Simulation and Monte Carlo: With Applications in Finance and
  MCMC
DANIEL · Applications of Statistics to Industrial Experimentation
DANIEL · Biostatistics: A Foundation for Analysis in the Health Sciences, *Eighth
  Edition*
* DANIEL · Fitting Equations to Data: Computer Analysis of Multifactor Data,
  Second Edition
DASU and JOHNSON · Exploratory Data Mining and Data Cleaning
DAVID and NAGARAJA · Order Statistics, *Third Edition*
DAVINO, FURNO and VISTOCCO · Quantile Regression: Theory and
  Applications
* DEGROOT, FIENBERG, and KADANE · Statistics and the Law
DEL CASTILLO · Statistical Process Adjustment for Quality Control
DeMARIS · Regression with Social Data: Modeling Continuous and Limited
  Response Variables
DEMIDENKO · Mixed Models: Theory and Applications with R, *Second Edition*
DENISON, HOLMES, MALLICK and SMITH · Bayesian Methods for Nonlinear
  Classification and Regression
DETTE and STUDDEN · The Theory of Canonical Moments with Applications in
  Statistics, Probability, and Analysis

*Now available in a lower priced paperback edition in the Wiley Classics Library.
†Now available in a lower priced paperback edition in the Wiley–Interscience Paperback
Series.

DEY and MUKERJEE · Fractional Factorial Plans
DILLON and GOLDSTEIN · Multivariate Analysis: Methods and Applications
* DODGE and ROMIG · Sampling Inspection Tables, *Second Edition*
* DOOB · Stochastic Processes
DOWDY, WEARDEN, and CHILKO · Statistics for Research, *Third Edition*
DRAPER and SMITH · Applied Regression Analysis, *Third Edition*
DRYDEN and MARDIA · Statistical Shape Analysis
DUDEWICZ and MISHRA · Modern Mathematical Statistics
DUNN and CLARK · Basic Statistics: A Primer for the Biomedical Sciences, *Fourth Edition*
DUPUIS and ELLIS · A Weak Convergence Approach to the Theory of Large Deviations
EDLER and KITSOS · Recent Advances in Quantitative Methods in Cancer and Human Health Risk Assessment
* ELANDT-JOHNSON and JOHNSON · Survival Models and Data Analysis
ENDERS · Applied Econometric Time Series, *Third Edition*
† ETHIER and KURTZ · Markov Processes: Characterization and Convergence
EVANS, HASTINGS, and PEACOCK · Statistical Distributions, *Third Edition*
EVERITT, LANDAU, LEESE, and STAHL · Cluster Analysis, *Fifth Edition*
FEDERER and KING · Variations on Split Plot and Split Block Experiment Designs
FELLER · An Introduction to Probability Theory and Its Applications, Volume I, *Third Edition,* Revised; Volume II, *Second Edition*
FITZMAURICE, LAIRD, and WARE · Applied Longitudinal Analysis, *Second Edition*
* FLEISS · The Design and Analysis of Clinical Experiments
FLEISS · Statistical Methods for Rates and Proportions, *Third Edition*
† FLEMING and HARRINGTON · Counting Processes and Survival Analysis
FUJIKOSHI, ULYANOV, and SHIMIZU · Multivariate Statistics: High-Dimensional and Large-Sample Approximations
FULLER · Introduction to Statistical Time Series, *Second Edition*
† FULLER · Measurement Error Models
GALLANT · Nonlinear Statistical Models
GEISSER · Modes of Parametric Statistical Inference
GELMAN and MENG · Applied Bayesian Modeling and Causal Inference from ncomplete-Data Perspectives
GEWEKE · Contemporary Bayesian Econometrics and Statistics
GHOSH, MUKHOPADHYAY, and SEN · Sequential Estimation
GIESBRECHT and GUMPERTZ · Planning, Construction, and Statistical Analysis of Comparative Experiments
GIFI · Nonlinear Multivariate Analysis
GIVENS and HOETING · Computational Statistics
GLASSERMAN and YAO · Monotone Structure in Discrete-Event Systems
GNANADESIKAN · Methods for Statistical Data Analysis of Multivariate Observations, *Second Edition*
GOLDSTEIN · Multilevel Statistical Models, *Fourth Edition*
GOLDSTEIN and LEWIS · Assessment: Problems, Development, and Statistical Issues
GOLDSTEIN and WOOFF · Bayes Linear Statistics
GRAHAM · Markov Chains: Analytic and Monte Carlo Computations
GREENWOOD and NIKULIN · A Guide to Chi-Squared Testing
GROSS, SHORTLE, THOMPSON, and HARRIS · Fundamentals of Queueing Theory, *Fourth Edition*

---

*Now available in a lower priced paperback edition in the Wiley Classics Library.
†Now available in a lower priced paperback edition in the Wiley–Interscience Paperback Series.

GROSS, SHORTLE, THOMPSON, and HARRIS · Solutions Manual to
  Accompany Fundamentals of Queueing Theory, *Fourth Edition*
* HAHN and SHAPIRO · Statistical Models in Engineering
  HAHN and MEEKER · Statistical Intervals: A Guide for Practitioners
  HALD · A History of Probability and Statistics and their Applications Before 1750
† HAMPEL · Robust Statistics: The Approach Based on Influence Functions
  HARTUNG, KNAPP, and SINHA · Statistical Meta-Analysis with Applications
  HEIBERGER · Computation for the Analysis of Designed Experiments
  HEDAYAT and SINHA · Design and Inference in Finite Population Sampling
  HEDEKER and GIBBONS · Longitudinal Data Analysis
  HELLER · MACSYMA for Statisticians
  HERITIER, CANTONI, COPT, and VICTORIA-FESER · Robust Methods in
  Biostatistics
  HINKELMANN and KEMPTHORNE · Design and Analysis of Experiments,
  Volume 1: Introduction to Experimental Design, *Second Edition*
  HINKELMANN and KEMPTHORNE · Design and Analysis of Experiments,
  Volume 2: Advanced Experimental Design
  HINKELMANN (editor) · Design and Analysis of Experiments, Volume 3: Special
  Designs and Applications
  HOAGLIN, MOSTELLER, and TUKEY · Fundamentals of Exploratory Analysis
  of Variance
* HOAGLIN, MOSTELLER, and TUKEY · Exploring Data Tables, Trends and
  Shapes
* HOAGLIN, MOSTELLER, and TUKEY · Understanding Robust and Exploratory
  Data Analysis
  HOCHBERG and TAMHANE · Multiple Comparison Procedures
  HOCKING · Methods and Applications of Linear Models: Regression and the
  Analysis of Variance, *Third Edition*
  HOEL · Introduction to Mathematical Statistics, *Fifth Edition*
  HOGG and KLUGMAN · Loss Distributions
  HOLLANDER, WOLFE, and CHICKEN · Nonparametric Statistical Methods,
  Third Edition
  HOSMER and LEMESHOW · Applied Logistic Regression, *Second Edition*
  HOSMER, LEMESHOW, and MAY · Applied Survival Analysis: Regression
  Modeling of Time-to-Event Data, *Second Edition*
  HUBER · Data Analysis: What Can Be Learned From the Past 50 Years
  HUBER · Robust Statistics
† HUBER and RONCHETTI · Robust Statistics, *Second Edition*
  HUBERTY · Applied Discriminant Analysis, *Second Edition*
  HUBERTY and OLEJNIK · Applied MANOVA and Discriminant Analysis, *Second
  Edition*
  HUITEMA · The Analysis of Covariance and Alternatives: Statistical Methods for
  Experiments, Quasi-Experiments, and Single-Case Studies, *Second Edition*
  HUNT and KENNEDY · Financial Derivatives in Theory and Practice, *Revised
  Edition*
  HURD and MIAMEE · Periodically Correlated Random Sequences: Spectral
  Theory and Practice
  HUSKOVA, BERAN, and DUPAC · Collected Works of Jaroslav Hajek— with
  Commentary
  HUZURBAZAR · Flowgraph Models for Multistate Time-to-Event Data
  JACKMAN · Bayesian Analysis for the Social Sciences
† JACKSON · A User's Guide to Principle Components
  JOHN · Statistical Methods in Engineering and Quality Assurance
  JOHNSON · Multivariate Statistical Simulation

---

*Now available in a lower priced paperback edition in the Wiley Classics Library.
†Now available in a lower priced paperback edition in the Wiley–Interscience Paperback
Series.

JOHNSON and BALAKRISHNAN · Advances in the Theory and Practice of Statistics: A Volume in Honor of Samuel Kotz

JOHNSON, KEMP, and KOTZ · Univariate Discrete Distributions, *Third Edition*

JOHNSON and KOTZ (editors) · Leading Personalities in Statistical Sciences: From the Seventeenth Century to the Present

JOHNSON, KOTZ, and BALAKRISHNAN · Continuous Univariate Distributions, Volume 1, *Second Edition*

JOHNSON, KOTZ, and BALAKRISHNAN · Continuous Univariate Distributions, Volume 2, *Second Edition*

JOHNSON, KOTZ, and BALAKRISHNAN · Discrete Multivariate Distributions

JUDGE, GRIFFITHS, HILL, LÜTKEPOHL, and LEE · The Theory and Practice of Econometrics, *Second Edition*

JUREK and MASON · Operator-Limit Distributions in Probability Theory

KADANE · Bayesian Methods and Ethics in a Clinical Trial Design

KADANE AND SCHUM · A Probabilistic Analysis of the Sacco and Vanzetti Evidence

KALBFLEISCH and PRENTICE · The Statistical Analysis of Failure Time Data, *Second Edition*

KARIYA and KURATA · Generalized Least Squares

KASS and VOS · Geometrical Foundations of Asymptotic Inference

† KAUFMAN and ROUSSEEUW · Finding Groups in Data: An Introduction to Cluster Analysis

KEDEM and FOKIANOS · Regression Models for Time Series Analysis

KENDALL, BARDEN, CARNE, and LE · Shape and Shape Theory

KHURI · Advanced Calculus with Applications in Statistics, *Second Edition*

KHURI, MATHEW, and SINHA · Statistical Tests for Mixed Linear Models

* KISH · Statistical Design for Research

KLEIBER and KOTZ · Statistical Size Distributions in Economics and Actuarial Sciences

KLEMELÄ · Smoothing of Multivariate Data: Density Estimation and Visualization

KLUGMAN, PANJER, and WILLMOT · Loss Models: From Data to Decisions, *Third Edition*

KLUGMAN, PANJER, and WILLMOT · Loss Models: Further Topics

KLUGMAN, PANJER, and WILLMOT · Solutions Manual to Accompany Loss Models: From Data to Decisions, *Third Edition*

KOSKI and NOBLE · Bayesian Networks: An Introduction

KOTZ, BALAKRISHNAN, and JOHNSON · Continuous Multivariate Distributions, Volume 1, *Second Edition*

KOTZ and JOHNSON (editors) · Encyclopedia of Statistical Sciences: Volumes 1 to 9 with Index

KOTZ and JOHNSON (editors) · Encyclopedia of Statistical Sciences: Supplement Volume

KOTZ, READ, and BANKS (editors) · Encyclopedia of Statistical Sciences: Update Volume 1

KOTZ, READ, and BANKS (editors) · Encyclopedia of Statistical Sciences: Update Volume 2

KOWALSKI and TU · Modern Applied U-Statistics

KRISHNAMOORTHY and MATHEW · Statistical Tolerance Regions: Theory, Applications, and Computation

KROESE, TAIMRE, and BOTEV · Handbook of Monte Carlo Methods

KROONENBERG · Applied Multiway Data Analysis

KULINSKAYA, MORGENTHALER, and STAUDTE · Meta Analysis: A Guide to Calibrating and Combining Statistical Evidence

*Now available in a lower priced paperback edition in the Wiley Classics Library.
†Now available in a lower priced paperback edition in the Wiley–Interscience Paperback Series.

*Now available in a lower priced paperback edition in the Wiley Classics Library.
†Now available in a lower priced paperback edition in the Wiley–Interscience Paperback
Series.

RENCHER and SCHAALJE · Linear Models in Statistics, *Second Edition*
RENCHER and CHRISTENSEN · Methods of Multivariate Analysis, *Third Edition*
RENCHER · Multivariate Statistical Inference with Applications
RIGDON and BASU · Statistical Methods for the Reliability of Repairable Systems
* RIPLEY · Spatial Statistics
* RIPLEY · Stochastic Simulation
ROHATGI and SALEH · An Introduction to Probability and Statistics, *Second Edition*
ROLSKI, SCHMIDLI, SCHMIDT, and TEUGELS · Stochastic Processes for Insurance and Finance
ROSENBERGER and LACHIN · Randomization in Clinical Trials: Theory and Practice
ROSSI, ALLENBY, and McCULLOCH · Bayesian Statistics and Marketing
† ROUSSEEUW and LEROY · Robust Regression and Outlier Detection
ROYSTON and SAUERBREI · Multivariate Model Building: A Pragmatic Approach to Regression Analysis Based on Fractional Polynomials for Modeling Continuous Variables
* RUBIN · Multiple Imputation for Nonresponse in Surveys
RUBINSTEIN and KROESE · Simulation and the Monte Carlo Method, *Second Edition*
RUBINSTEIN and MELAMED · Modern Simulation and Modeling
RUBINSTEIN, RIDDER, and VAISMAN · Fast Sequential Monte Carlo Methods for Counting and Optimization
RYAN · Modern Engineering Statistics
RYAN · Modern Experimental Design
RYAN · Modern Regression Methods, *Second Edition*
RYAN · Sample Size Determination and Power
RYAN · Statistical Methods for Quality Improvement, *Third Edition*
SALEH · Theory of Preliminary Test and Stein-Type Estimation with Applications
SALTELLI, CHAN, and SCOTT (editors) · Sensitivity Analysis
SCHERER · Batch Effects and Noise in Microarray Experiments: Sources and Solutions
* SCHEFFE · The Analysis of Variance
SCHIMEK · Smoothing and Regression: Approaches, Computation, and Application
SCHOTT · Matrix Analysis for Statistics, *Second Edition*
SCHOUTENS · Levy Processes in Finance: Pricing Financial Derivatives
SCOTT · Multivariate Density Estimation: Theory, Practice, and Visualization
* SEARLE · Linear Models
† SEARLE · Linear Models for Unbalanced Data
† SEARLE · Matrix Algebra Useful for Statistics
† SEARLE, CASELLA, and McCULLOCH · Variance Components
SEARLE and WILLETT · Matrix Algebra for Applied Economics
SEBER · A Matrix Handbook For Statisticians
† SEBER · Multivariate Observations
SEBER and LEE · Linear Regression Analysis, *Second Edition*
† SEBER and WILD · Nonlinear Regression
SENNOTT · Stochastic Dynamic Programming and the Control of Queueing Systems
* SERFLING · Approximation Theorems of Mathematical Statistics
SHAFER and VOVK · Probability and Finance: It's Only a Game!
SHERMAN · Spatial Statistics and Spatio-Temporal Data: Covariance Functions and Directional Properties

*Now available in a lower priced paperback edition in the Wiley Classics Library.
†Now available in a lower priced paperback edition in the Wiley–Interscience Paperback Series.

WINKER · Optimization Heuristics in Economics: Applications of Threshold Accepting

WOODWORTH · Biostatistics: A Bayesian Introduction

WOOLSON and CLARKE · Statistical Methods for the Analysis of Biomedical Data, *Second Edition*

WU and HAMADA · Experiments: Planning, Analysis, and Parameter Design Optimization, *Second Edition*

WU and ZHANG · Nonparametric Regression Methods for Longitudinal Data Analysis

YAKIR · Extremes in Random Fields

YIN · Clinical Trial Design: Bayesian and Frequentist Adaptive Methods

YOUNG, VALERO-MORA, and FRIENDLY · Visual Statistics: Seeing Data with Dynamic Interactive Graphics

ZACKS · Examples and Problems in Mathematical Statistics

ZACKS · Stage-Wise Adaptive Designs

* ZELLNER · An Introduction to Bayesian Inference in Econometrics

ZELTERMAN · Discrete Distributions—Applications in the Health Sciences

ZHOU, OBUCHOWSKI, and McCLISH · Statistical Methods in Diagnostic Medicine, *Second Edition*